Environmental Transport Phenomena

GREEN CHEMISTRY AND CHEMICAL ENGINEERING

Series Editor: Sunggyu Lee
Ohio University, Athens, Ohio, USA

Proton Exchange Membrane Fuel Cells: Contamination and Mitigation Strategies
Hui Li, Shanna Knights, Zheng Shi, John W. Van Zee, and Jiujun Zhang

Proton Exchange Membrane Fuel Cells: Materials Properties and Performance
David P. Wilkinson, Jiujun Zhang, Rob Hui, Jeffrey Fergus, and Xianguo Li

Solid Oxide Fuel Cells: Materials Properties and Performance
Jeffrey Fergus, Rob Hui, Xianguo Li, David P. Wilkinson, and Jiujun Zhang

Efficiency and Sustainability in the Energy and Chemical Industries: Scientific Principles and Case Studies, Second Edition
Krishnan Sankaranarayanan, Jakob de Swaan Arons, and Hedzer van der Kooi

Nuclear Hydrogen Production Handbook
Xing L. Yan and Ryutaro Hino

Magneto Luminous Chemical Vapor Deposition
Hirotsugu Yasuda

Carbon-Neutral Fuels and Energy Carriers
Nazim Z. Muradov and T. Nejat Veziroğlu

Oxide Semiconductors for Solar Energy Conversion: Titanium Dioxide
Janusz Nowotny

Lithium-Ion Batteries: Advanced Materials and Technologies
Xianxia Yuan, Hansan Liu, and Jiujun Zhang

Process Integration for Resource Conservation
Dominic C. Y. Foo

Chemicals from Biomass: Integrating Bioprocesses into Chemical Production Complexes for Sustainable Development
Debalina Sengupta and Ralph W. Pike

Hydrogen Safety
Fotis Rigas and Paul Amyotte

Biofuels and Bioenergy: Processes and Technologies
Sunggyu Lee and Y. T. Shah

Hydrogen Energy and Vehicle Systems
Scott E. Grasman

Integrated Biorefineries: Design, Analysis, and Optimization
Paul R. Stuart and Mahmoud M. El-Halwagi

Water for Energy and Fuel Production
Yatish T. Shah

Handbook of Alternative Fuel Technologies, Second Edition
Sunggyu Lee, James G. Speight, and Sudarshan K. Loyalka

Environmental Transport Phenomena
A. Eduardo Sáez and James C. Baygents

Environmental Transport Phenomena

A. Eduardo Sáez
James C. Baygents

CRC Press
Taylor & Francis Group
Boca Raton London New York

CRC Press is an imprint of the
Taylor & Francis Group, an **informa** business

CRC Press
Taylor & Francis Group
6000 Broken Sound Parkway NW, Suite 300
Boca Raton, FL 33487-2742

Printed on acid-free paper
Version Date: 20140916

International Standard Book Number-13: 978-1-4665-7623-0 (Hardback)

Library of Congress Cataloging-in-Publication Data

Sáez, A. Eduardo.
Environmental transport phenomena / A. Eduardo Sáez and James C. Baygents.
pages cm. -- (Green chemistry and chemical engineering)
Includes bibliographical references and index.
ISBN 978-1-4665-7623-0 (hardback)
1. Cross-media pollution. 2. Environmental chemistry. 3. Diffusion. I. Baygents, James Claude. II. Title.

TD193.3S24 2014
577.27--dc23 2014036258

Visit the Taylor & Francis Web site at
http://www.taylorandfrancis.com

and the CRC Press Web site at
http://www.crcpress.com

Contents

Series Preface

The subjects and disciplines of chemistry and chemical engineering have encountered a new landmark in the way of thinking about, developing, and designing chemical products and processes. This revolutionary philosophy, termed "green chemistry and chemical engineering," focuses on the designs of products and processes that are conducive to reducing or eliminating the use and generation of hazardous substances. In dealing with hazardous or potentially hazardous substances, there may be some overlaps and interrelationships between environmental chemistry and green chemistry. While environmental chemistry is the chemistry of the natural environment and the pollutant chemicals in nature, green chemistry proactively aims to reduce and prevent pollution at its very source. In essence, the philosophies of green chemistry and chemical engineering tend to focus more on industrial application and practice rather than academic principles and phenomenological science. However, as both chemistry and chemical engineering philosophy, green chemistry and chemical engineering derives from and builds upon organic chemistry, inorganic chemistry, polymer chemistry, fuel chemistry, biochemistry, analytical chemistry, physical chemistry, environmental chemistry, thermodynamics, chemical reaction engineering, transport phenomena, chemical process design, separation technology, automatic process control, and more. In short, green chemistry and chemical engineering is the rigorous use of chemistry and chemical engineering for pollution prevention and environmental protection.

The Pollution Prevention Act of 1990 in the United States established a national policy to prevent or reduce pollution at its source whenever feasible. And adhering to the spirit of this policy, the Environmental Protection Agency (EPA) launched its Green Chemistry Program to promote innovative chemical technologies that reduce or eliminate the use or generation of hazardous substances in the design, manufacture, and use of chemical products. The global efforts in green chemistry and chemical engineering have recently gained a substantial amount of support from the international community of science, engineering, academia, industry, and governments in all phases and aspects. Some of the successful examples and key technological developments include the use of supercritical carbon dioxide as green solvent in separation technologies, application of supercritical water oxidation for destruction of harmful substances, process integration with carbon dioxide sequestration steps, solvent-free synthesis of chemicals and polymeric materials, exploitation of biologically degradable materials, use of aqueous hydrogen peroxide for efficient oxidation, development of hydrogen proton exchange membrane (PEM) fuel cells for a variety of power generation needs, advanced biofuel productions, devulcanization of spent tire rubber, avoidance of the use of chemicals and processes causing generation of volatile organic compounds (VOCs), replacement of traditional petrochemical processes by microorganism-based bioengineering processes, replacement of chlorofluorocarbons (CFCs) with nonhazardous alternatives, advances in design of energy-efficient processes, use of clean, alternative and renewable energy sources in manufacturing, and much more. This list, even though it is only a partial compilation, is undoubtedly growing exponentially.

This book series on Green Chemistry and Chemical Engineering by CRC Press/Taylor & Francis is designed to meet the new challenges of the twenty-first century in the chemistry and chemical engineering disciplines by publishing books and monographs based on cutting-edge research and development to the effect of reducing adverse impacts on the environment by chemical enterprise. And in achieving this, the series will detail the development of alternative sustainable technologies that will minimize the hazard and maximize the efficiency of any chemical choice. The series aims at delivering the readers in academia and industry with an authoritative information source in the field of green chemistry and chemical engineering. The publisher and its series editor are fully aware of the rapidly evolving nature of the subject and its long-lasting impact on the quality of human life in both the present and future. As such, the team is committed to making this series the most comprehensive and accurate literary source in the field of green chemistry and chemical engineering.

Sunggyu Lee

Preface

The study of transport phenomena has been one of the main subjects of chemical engineering since the introduction of the homonymous book by Bird, Stewart, and Lightfoot in the 1960s, which was one of the first attempts to unify the disciplines of fluid mechanics, heat transfer, and mass transfer into a single topic. Nowadays, applications of transport phenomena are ubiquitous in science and engineering and, in particular, in the study of environmental processes, both natural and engineered.

Our main aim in writing this book was to deliver an introduction of transport phenomena to environmental engineers and scientists. Specifically, this book treats in detail the subjects of mass and momentum transport, placing special emphasis on applications to environmental processes. We build on the fundamental principles of mass and momentum conservation to develop a mathematical framework to formulate and to solve transport phenomena problems. As such, this work may also serve as an introduction to the topic for chemical engineers interested in applications of transport phenomena to environmental systems.

This textbook is intended for use in a one-semester undergraduate or graduate introductory course in transport phenomena. One of our objectives has been to present the material in a way that is accessible to students of chemical and environmental engineering. It is expected that readers have been exposed to an introductory course on ordinary differential equations. In general, we present mathematical derivations in detail to provide the reader with tools that can be used to solve other problems. Familiarity with basic applications of fluid mechanics is also desirable.

The idea of writing this book originated in conversations between the authors and Professor Robert G. Arnold. We are grateful to Bob for his continuous encouragement and support for the project, without which it never would had come to fruition. We are thankful to Professor James Farrell for adopting earlier versions of our manuscript to teach the course *Environmental Transport Phenomena* at the University of Arizona. Our gratitude also goes to the many hardworking and talented students who have been instrumental in shaping our approach to teach the subject.

A. Eduardo Sáez
James C. Baygents
Tucson, September 2014

Authors

A. Eduardo Sáez is a distinguished professor at the University of Arizona (UA) Department of Chemical and Environmental Engineering and an adjunct professor at the Mel and Enid Zuckerman College of Public Health at the University of Arizona. He is from Venezuela, where he earned a BSc in chemical engineering at Simón Bolívar University. Subsequently, he earned an MS and a PhD in chemical engineering at the University of California at Davis.

He was a faculty member at Simón Bolívar University and North Carolina State University before joining UA in 1998. He has been a visiting scholar at the Department of Physics at Bristol University (UK) and at Pacific Northwest National Laboratory in Richland, Washington. Dr. Sáez has earned numerous awards for research and teaching.

He is part of a legion of increasingly relevant engineers confronting environmental challenges. His research deals with the fate, transport, and treatment of trace contaminants in water sources, including the use of solar energy to drive advanced oxidation of contaminants in reuse water resources. He is also interested in the transport of metal and metalloid contaminants in windblown dust from mining operations in arid and semi-arid environments. This research includes monitoring, analyzing, and modeling arsenic and lead transport in mining field sites in Arizona. More recently, the research group has expanded to join an international effort to study contaminant transport in mining sites in Australia.

James C. Baygents is the associate dean for academic affairs at the College of Engineering at the University of Arizona. He is a member of the Department of Chemical and Environmental Engineering (ChEE) and the Program in Applied Mathematics at UA.

Dr. Baygents joined the UA engineering faculty as an assistant professor in 1991, the same year he earned a PhD in chemical engineering at Princeton University. He also holds an MA (Princeton, 1981) and a BS (Rice, 1980) in chemical engineering. For three years prior to joining the UA, Dr. Baygents was a visiting scientist, then a research fellow, at the Space Science Laboratory at the NASA Marshall Space Flight Center in Huntsville, Alabama. In 1995, he received the Arizona Mortar Board Senior Honor Society award for outstanding faculty service. In 1997, he was awarded an International Research Fellowship by the National Science Foundation for study at the University of Melbourne. In 2009, he was recognized by ChEE and the College for Excellence at the Student Interface. He is a member of Phi Beta Kappa, Tau Beta Pi, and Phi Lambda Upsilon honor societies, as well as the College of Fellows at Rice University's Will Rice College.

Dr. Baygents's research interests include transport processes in natural and engineered systems; separations and water treatment processes; diffusion–reaction–precipitation in aqueous electrolyte systems; electrokinetic theory, measurements, and separations; electrically driven fluid motion and transport processes, including microfluidics; pattern formation in caves associated with Karst water systems; and industrial water treatment for recycle and reuse.

1 Introduction

1.1 SCOPE AND MOTIVATION

This book deals with mass transport in systems that are relevant to environmental engineering applications. The phrase "mass transport" refers to the motion and spatial distribution of a chemical species in a multicomponent mixture. In general, the chemical species of interest will be a contaminant, and the environmental system may involve the presence of more than one phase.

Environmental and chemical engineers are interested in mass transport mainly because of three important objectives:

1. The design of engineered systems. Removal and treatment of contaminants require quantification of mass transport rates. Separations and reactive processes are examples of practical cases in which a contaminant must be transported through a mixture.
2. The understanding of natural systems. The distribution and fate of contaminants in natural systems is usually dictated by mass transport. Here, we may be concerned with processes that span a wide variety of scales: from transport of a compound through a cell wall to the distribution of greenhouse gases in the atmosphere.
3. Transport modeling. The ability to predict the spatial and temporal distribution of a contaminant in both engineered and natural systems is one of the main goals of the study of mass transport.

To accomplish the objectives mentioned, we have the following tools at our disposal:

1. Conservation principles. These are physical laws that, as such, are independent of the scale of observation and are applied to the continuum. Conservation of mass, momentum, and energy are the basic principles employed in transport applications.
2. Thermodynamic relations. These are properties and equations that describe a system at equilibrium. They include P–V–T relations for a pure substance (e.g., the ideal gas equation), thermal properties (e.g., relation between internal energy and temperature), the distribution of a component between two phases at equilibrium (e.g., Henry's law for gas–liquid equilibrium).
3. Constitutive equations and material behavior. These represent a link between conservation principles in the continuum and the molecular scale.

For example, the equations of Newton (momentum), Fourier (energy), and Fick (mass) represent the behavior of materials due to molecular-level interactions. Another important example is the kinetic rate equations for chemical reactions.

4. Mathematical techniques. The mathematical complexity of the transport equations often requires the capability to solve ordinary and partial differential equations. Analytical methods are usually preferred, but numerical methods are commonly required.

The book is organized as follows:

Chapter 2 presents fundamentals of the application of the principle of mass conservation to perform mass balances in single-component systems and in multicomponent systems that are well mixed; that is, systems in which rates of species mass transfer are so fast that concentrations may be considered spatially uniform. These fundamentals can be used in the design of well-mixed chemical reactors and to estimate temporal concentration variations in natural systems.

Chapter 3 introduces diffusion as a mechanism of mass transfer. The objective of this chapter is to provide the tools necessary to quantify diffusion as a continuum mass transport mechanism, so that simple practical problems in which diffusion prevails can be formulated and solved.

The fundamentals presented in Chapters 2 and 3 are generalized in Chapter 4, where the general point equations that represent mass transfer in a multicomponent system are derived and used to solve diffusion problems of practical relevance. In this chapter, the scope is widened to include relatively complex problems that require advanced mathematical methods of solution.

Chapter 5 deals with mass transfer in the presence of chemical reactions. Reactive systems are ubiquitous in both natural and engineered systems and often mass transfer becomes the limiting rate step that controls the conversion of a substance in a chemical reaction, which limits its productivity. For example, the rate at which a contaminant is destroyed by a technique involving chemical transformations might be a result of the combination of the kinetics of the reactions that take place and the speed at which the reactants are transferred through space.

Most applications in Chapters 3 through 5 deal with static systems in which diffusion is the prevalent mass transport mechanism. Chapter 6 deals with mass transfer in moving fluids, in which transport due to the bulk motion of the mixture, termed convection or advection, is an important mechanism. First, the use of mass and momentum balances are introduced as tolls that can be used to predict the velocity field in a fluid in motion. Later, these concepts are applied to describe in a comprehensive manner mass transport in moving fluids.

The combination of convection and diffusion is explored quantitatively in Chapter 7 where the concept of hydrodynamic dispersion is introduced and quantified. Understanding dispersion as a result of combined diffusion and convection is particularly important in the study of the distribution of contaminants in natural systems, as well as in the description of separation processes such as adsorption, ion exchange, and chromatography.

Environmental engineers and scientists often deal with multiphase systems. Transport of contaminants in soils, aquifers, and engineered systems that involve separation of a species from a fluid to a solid, such as adsorption, can be understood when the fundamentals of mass transfer are applied in the context of multiphase systems, which is the theme of Chapter 8.

2 Mass Conservation and Macroscopic Mass Balances

2.1 THE PRINCIPLE OF CONSERVATION OF MASS AND THE TOTAL MASS BALANCE

The most important physical law relevant to mass transport processes is the principle of conservation of mass. This principle will be applied to (i) the conservation of the total mass of a material, and (ii) the mass balance of a specific chemical species in a mixture. In this section, we will derive and apply total mass balance equations for arbitrary control volumes. The resulting equations are called **macroscopic mass balances** and they are useful to solve problems in physical domains that are well mixed; that is, where species concentrations are spatially uniform.

The basis of application of the macroscopic mass balance is the definition of a control volume. We will term this control volume **arbitrary volume**, V_a, which is a region of space defined by the observer (Figure 2.1). The arbitrary volume is defined by specifying the surface that delimits it, A_a. In general, the arbitrary volume may change with time, and it may move through space. Points in the arbitrary volume move with a velocity field **w**, which, in general, is a function of position and time. We will use bold symbols to represent vectors. A brief review of vector algebra is presented in Appendix A. It is important to note that V_a, A_a, and **w** are defined by the observer as a basis for the analysis of a specific system.

The principle of conservation of mass applied to an arbitrary volume is formulated here as an instantaneous mass balance, that is, the result of mass conservation at a given time:

$$\underset{\text{(accumulation)}}{\left\{\begin{array}{l}\text{The time rate of change of}\\ \text{the total mass in an}\\ \text{arbitrary volume}\end{array}\right\}} = \underset{\text{(input)}}{\left\{\begin{array}{l}\text{Mass flow rate entering}\\ \text{the arbitrary volume}\\ \text{through its surface}\end{array}\right\}} - \underset{\text{(output)}}{\left\{\begin{array}{l}\text{Mass flow rate leaving}\\ \text{the arbitrary volume}\\ \text{through its surface}\end{array}\right\}} \tag{2.1}$$

This equation will be formulated now in terms of its most general mathematical statement, using **point or intensive properties** of the material. First, consider the left-hand side of the equation. The total mass contained in the arbitrary volume

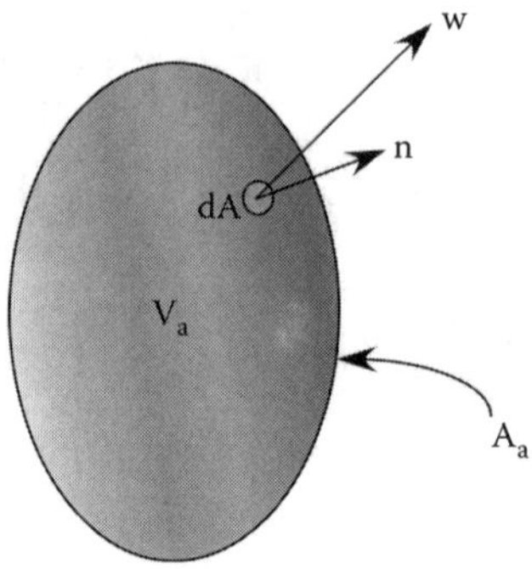

FIGURE 2.1 An arbitrary volume, V_a, delimited by an arbitrary surface, A_a. At each point on the surface surrounded by an area differential dA, the normal unit vector, **n**, pointing outwards is shown. The velocity with which points on the arbitrary surface move is **w**, which can be, in general, a function of position and time.

can be expressed as the volume integral of the density (mass per unit volume) of the material inside V_a:

$$\begin{Bmatrix}\text{Total mass in the}\\ \text{arbitrary volume}\end{Bmatrix} = M = \int_{V_a(t)} \rho dV \tag{2.2}$$

where we are explicitly considering the general case in which V_a is a function of time. The left-hand side of Equation 2.1 is

$$\begin{Bmatrix}\text{The time rate of change of}\\ \text{the total mass in the}\\ \text{arbitrary volume}\end{Bmatrix} = \frac{dM}{dt} = \frac{d}{dt}\int_{V_a(t)} \rho dV \tag{2.3}$$

The right-hand side of Equation 2.1 will be nonzero only when material crosses $A_a(t)$. Let $\mathbf{v}_r$ be the velocity vector of the material relative to the surface. The material entering or leaving the arbitrary volume will be doing so with a velocity given by the component of the vector $\mathbf{v}_r$ in the direction normal to the surface, $(\mathbf{v}_r)_n$ (Figure 2.2). The normal component of $\mathbf{v}_r$ can be obtained by the scalar product between the vectors $\mathbf{v}_r$ and **n**:

$$(\mathbf{v}_r)_n = \mathbf{v}_r \cdot \mathbf{n} \tag{2.4}$$

Note that, since **n** points outwards, $\mathbf{v}_r \cdot \mathbf{n} > 0$ means that material **leaves** the arbitrary volume through dA whereas $\mathbf{v}_r \cdot \mathbf{n} < 0$ means that material **enters** the arbitrary volume through dA. In view of these definitions, we have:

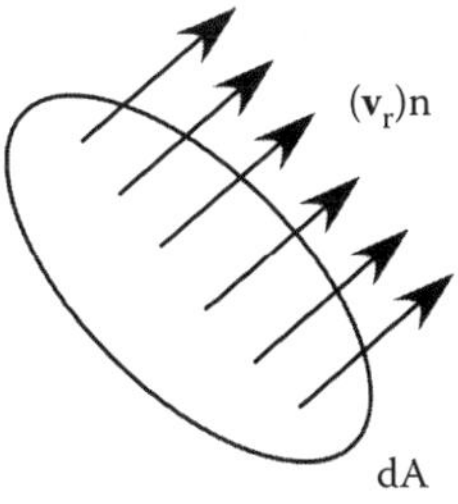

FIGURE 2.2 Material crossing the arbitrary surface: $(\mathbf{v}_r)_n$ is the normal component of the relative velocity of the material with respect to the surface.

$$\begin{Bmatrix}\text{Volume of material that}\\ \text{crosses dA per unit time}\end{Bmatrix} = \mathbf{v}_r \cdot \mathbf{n}dA \tag{2.5}$$

and, therefore,

$$\begin{Bmatrix}\text{Mass of material that}\\ \text{crosses dA per unit time}\end{Bmatrix} = \rho\mathbf{v}_r \cdot \mathbf{n}dA \tag{2.6}$$

The integral of Equation 2.6 over $A_a(t)$ will add all the mass that leaves the arbitrary volume per unit time and subtract all the mass that enters the arbitrary volume per unit time. Therefore, it will represent the **net** mass flow rate that **leaves** the arbitrary volume, which is the negative of the right-hand side in Equation 2.1. For this reason, we can state

$$\begin{Bmatrix}\text{Mass flow rate entering}\\ \text{the arbitrary volume}\\ \text{through its surface}\end{Bmatrix} - \begin{Bmatrix}\text{Mass flow rate leaving}\\ \text{the arbitrary volume}\\ \text{through its surface}\end{Bmatrix} = -\int_{A_a(t)} \rho\mathbf{v}_r \cdot \mathbf{n}dA \tag{2.7}$$

Combining Equations 2.1, 2.3, and 2.7, we find

$$\frac{d}{dt}\int_{V_{a(t)}} \rho dV = -\int_{A_{a(t)}} \rho\mathbf{v}_r \cdot \mathbf{n}dA \tag{2.8}$$

Let $\mathbf{v}$ be the velocity vector of the material at each point. The relative velocity of the material with respect to the surface is $\mathbf{v}_r = \mathbf{v} - \mathbf{w}$ and Equation 2.8 is expressed as

$$\frac{d}{dt}\int_{V_{a(t)}} \rho dV = \int_{A_{a(t)}} \rho\mathbf{n} \cdot (\mathbf{w} - \mathbf{v})dA \tag{2.9}$$

This equation is the **macroscopic mass balance**. It is usually written in terms of macroscopic (extensive) properties as follows:

$$\frac{dM}{dt} = \sum_{\text{all inlets } i} \dot{m}_i - \sum_{\text{all outlets } i} \dot{m}_i \tag{2.10}$$

where $\dot{m}_i$ represents mass flow rate through a specific section (inlet or outlet).

EXAMPLE 2.1 DRAINING OF A TANK

As an application of the macroscopic mass balance, consider the draining of liquid from a cylindrical tank of diameter D through an orifice of diameter d at its bottom. The liquid in the tank reaches a height H_0 at the beginning of the process $(t = 0)$ and, as the liquid drains, its height varies with time, $H = H(t)$. It is our objective to find this function.

To formulate the instantaneous mass balance for this problem, we can rely on the direct use of Equation 2.10, which can be regarded as a more intuitive application of the mass conservation principle. However, to illustrate the nomenclature employed in the derivation of the macroscopic mass balance, we will attempt to apply Equation 2.9 to the problem formulation.

Figure 2.3 shows how **v** and **w** vary over different portions of the surface. According to this, the integral on the right-hand side of Equation 2.9 simplifies to the integral over the outlet section, A_0,

$$\int_{A_{a(t)}} \rho \mathbf{n} \cdot (\mathbf{w} - \mathbf{v}) dA = -\int_{A_0} \rho \mathbf{n} \cdot \mathbf{v} dA \tag{2.11}$$

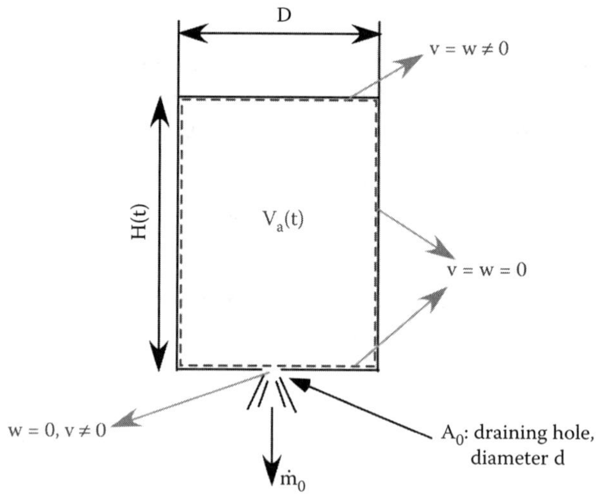

FIGURE 2.3 Draining of a cylindrical tank of diameter D through an orifice of diameter d. The arbitrary volume is the liquid contained inside the tank at a given time (volume enclosed by dashed lines).

Considering that the density is uniform and defining an average normal velocity on the outlet section by

$$v_0 = \frac{1}{A_0}\int_{A_0} \mathbf{n}\cdot\mathbf{v}dA \tag{2.12}$$

Equation 2.11 becomes

$$\int_{A_{a(t)}} \rho\mathbf{n}\cdot(\mathbf{w}-\mathbf{v})dA = -\rho v_0 A_0 \tag{2.13}$$

and the macroscopic mass balance (Equation 2.9) is

$$\frac{dM}{dt} = -\rho v_0 A_0 \tag{2.14}$$

Note that the mass flow rate through the orifice is $m_0 = \rho v_0 A_0$, so that Equation 2.14 is also the direct application of Equation 2.10.

Equation 2.14 can be rewritten by noting that $M = \rho V_a$. If the liquid density is constant, it simplifies from the equation, leading to

$$\frac{dV_a}{dt} = -v_0 A_0 \tag{2.15}$$

Now we use: $V_a = \pi D^2 H/4$ and $A_0 = \pi d^2/4$ to obtain

$$\frac{dH}{dt} = -\left(\frac{d}{D}\right)^2 v_0 \tag{2.16}$$

The drainage of the liquid occurs because of the gravitational force. Therefore, the higher the level of liquid in the tank the higher the weight of liquid, and the faster the exit velocity. Momentum conservation can be invoked to show that, in the absence of viscous effects, the exit velocity can be calculated from Bernoulli's equation (e.g., Bird et al., 2007, Chapter 3) to find

$$v_0 = \sqrt{2gH} \tag{2.17}$$

where g is the acceleration of gravity. For a typical orifice, this equation overestimates the exit velocity due to the occurrence of viscous friction. To take this into account, this equation is modified by introducing an empirical loss coefficient, $c < 1$, so that

$$v_0 = c\sqrt{2gH} \tag{2.18}$$

For a simple orifice in a flat surface, $c \approx 0.6$ (e.g., White, 1986, Chapter 6).

Using Equation 2.18, Equation 2.16 becomes

$$\frac{dH}{dt} = -\left(\frac{d}{D}\right)^2 c\sqrt{2gH} \tag{2.19}$$

This equation can be separated and integrated between an initial height of liquid H_0 ($t = 0$), and a generic time, as follows:

$$\int_{H_0}^{H} \frac{dH}{H^{1/2}} = -\left(\frac{d}{D}\right)^2 c\sqrt{2g}\int_0^t dt \tag{2.20}$$

which leads to

$$2\left(H^{1/2} - H_0^{1/2}\right) = -c\sqrt{2g}\left(\frac{d}{D}\right)^2 t \tag{2.21}$$

This equation can be rearranged as follows:

$$H = \left[H_0^{1/2} - c\sqrt{\frac{g}{2}}\left(\frac{d}{D}\right)^2 t\right]^2 \tag{2.22}$$

For example, for a tank with $D = 4.5$ m, $d = 2.5$ cm, $H_0 = 10$ m, using $c = 0.6$ and $g = 9.81$ m/s^2, Equation 2.22 yields

$$H = \left(3.162 - 4.1\times 10^{-5} t\right)^2, \text{t in s, H in m} \tag{2.23}$$

The time necessary to empty the tank completely is obtained by letting $H = 0$ in this equation, which yields: $t_f = 77{,}122$ s $= 21.4$ h. Figure 2.4 shows a plot of Equation 2.23, expressing the time in hours. The direct relation between height and time can be used to "mark" levels in the tank that correspond to specific times. This is the basis for a water clock (clepsydra), which was one of the first devices used in history to measure time.

EXAMPLE 2.2 ESCAPE OF HYDROGEN FROM A PRESSURIZED SYSTEM

A pressurized hydrogen (H_2) gas cylinder is used in an experiment in an environmental engineering laboratory (Figure 2.5). The cylinder has been incorrectly connected to the experiment by a short plastic tube. Hydrogen is such a small molecule that it can diffuse through the tube walls and escape to the environment. Since hydrogen is flammable, its leak is a fire hazard. Analysis of the hydrogen transport process through the tube walls has indicated that the mass flow rate of the hydrogen leak ($\dot{m}_L$) is directly proportional to the pressure difference across the tube wall,

$$\dot{m}_L = K(P_r - P_a) \tag{2.24}$$

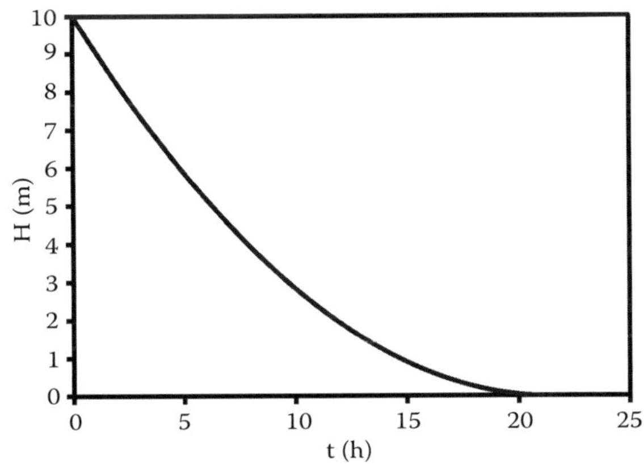

FIGURE 2.4 Height of liquid as a function of time in the draining of a tank.

where K is a known constant, and P_r and P_a are the pressures inside and outside the tube, respectively. Here, we will consider that the pressure inside the tube is approximately uniform and equal to the pressure after the regulating valve at all times.

Our objective in this problem is to predict how the tank pressure decreases with time due to the leak through the tube wall, that is, considering that the experiment is not consuming any hydrogen. The control volume for analysis is the gas contained inside the cylinder. Since the gas leaked from the tube will be the only outlet from the cylinder (considering that there is no accumulation in the tube itself), the macroscopic mass balance is

$$\frac{dM_T}{dt} = -\dot{m}_L = -K(P_r - P_a) \tag{2.25}$$

where M_T is the total mass of hydrogen in the tank at a given time. Considering that hydrogen behaves as an ideal gas, we can state

$$M_T = \frac{P_T V M}{RT} \tag{2.26}$$

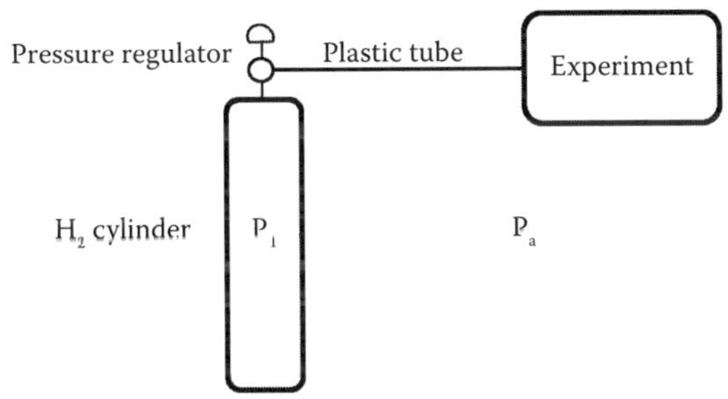

FIGURE 2.5 A pressurized hydrogen cylinder has been connected to an experiment by a plastic tube. The surroundings are at atmospheric pressure, P_a.

where V is the internal cylinder volume, M is the hydrogen molecular weight, R is the universal gas constant, and T the operating temperature. Substitution of Equation 2.26 into Equation 2.25 yields

$$\frac{dP_T}{dt} = -\frac{KRT}{VM}(P_r - P_a) \tag{2.27}$$

If the pressure regulator is working properly, P_r should be a constant, regardless of the value of P_T. This implies that the right-hand side of Equation 2.27 is constant and integration of Equation 2.27 from $t = 0$ (at which $PT = P_{T0}$) leads to the simple result

$$P_T = P_{T0} - \frac{KRT}{VM}(P_r - P_a)t \tag{2.28}$$

This result implies that the cylinder pressure decreases linearly with time. The flow will eventually stop when $P_T = P_r$.

Now, consider that the regulator malfunctions and the pressure inside the tube become equal to the pressure inside the cylinder at all times. In terms of our notation, this implies $P_r = P_T$, and Equation 2.27 now becomes

$$\frac{dP_T}{dt} = -\frac{KRT}{VM}(P_T - P_a) \tag{2.29}$$

In this case, to integrate the equation we must separate variables as follows:

$$\int_{P_{T0}}^{P_T} \frac{dP_T}{P_T - P_a} = -\frac{KRT}{VM}\int_0^t dt \tag{2.30}$$

which leads to, after integration and manipulation,

$$P_T = P_a + (P_{T0} - P_a)\exp\left(-\frac{KRT}{VM}t\right) \tag{2.31}$$

Note that, in this case, the pressure decreases exponentially until eventually ($t \to \infty$) is equalized with the atmospheric pressure, $P_T = P_a$.

2.2 MACROSCOPIC MASS BALANCE FOR A CHEMICAL SPECIES IN A MIXTURE

This section deals with the macroscopic mass balance for a chemical species in a multicomponent mixture. We will consider a generic component A in a mixture of N components ($A = 1, 2, \ldots, N$). One distinctive feature of this mass balance with respect to the total mass balance considered in the previous section is the possibility that species A is a reactant or a product in a chemical reaction.

The application of the principle of conservation of mass to species A in an arbitrary volume (Figure 2.1) is

$$\begin{Bmatrix}\text{The rate of change of}\\ \text{the mass of species A}\\ \text{in the arbitrary volume}\end{Bmatrix} = \begin{Bmatrix}\text{The rate at which mass of species A enters the}\\ \text{arbitrary volume through its boundaries}\end{Bmatrix} - \begin{Bmatrix}\text{The rate at which mass of species A leaves the}\\ \text{arbitrary volume through its boundaries}\end{Bmatrix} + \begin{Bmatrix}\text{The rate at which mass of species A is created}\\ \text{in the arbitrary volume by chemical reactions}\end{Bmatrix} - \begin{Bmatrix}\text{The rate at which mass of species A is consumed}\\ \text{in the arbitrary volume by chemical reactions}\end{Bmatrix} \quad (2.32)$$

The relative amount of species A at a point in the mixture is represented by a measure of its concentration. To establish the mathematical equivalent of Equation 2.32, we will use the **mass concentration** of species A, ρ_A (mass of A per unit mixture volume). Integration of the mass concentration over the arbitrary volume yields the total mass of A in it, and thus:

$$\begin{Bmatrix}\text{The rate of change of}\\ \text{the mass of species A}\\ \text{in the arbitrary volume}\end{Bmatrix} = \frac{d}{dt}\int_{V_a(t)} \rho_A dV \quad (2.33)$$

The first two terms on the right-hand side of Equation 2.32 can be quantified by an equation similar to Equation 2.7, with two main differences: ρ_A should be used instead of ρ, and the velocity $\mathbf{v}$ should be replaced by a species velocity, $\mathbf{v}_A$. We will discuss in more detail the meaning of the species velocity in Chapter 3. For the moment, we will simply state that each species does not move necessarily with the overall mixture velocity. We then have:

$$\begin{Bmatrix}\text{The rate at which mass of species A enters the}\\ \text{arbitrary volume through its boundaries}\end{Bmatrix} - \begin{Bmatrix}\text{The rate at which mass of species A leaves the}\\ \text{arbitrary volume through its boundaries}\end{Bmatrix} = \int_{A_a(t)} \rho_A \mathbf{n}\cdot(\mathbf{w}-\mathbf{v}_A)dA \quad (2.34)$$

Note that $\rho_A \mathbf{n}\cdot(\mathbf{w}-\mathbf{v}_A)dA$ represents the mass of A per unit time entering (when term is > 0) or leaving (when term is < 0) the arbitrary volume across the differential area element dA.

To quantify the reaction terms in Equation 2.32 we make use of the definition of a point or volumetric reaction rate, r_A, defined as the **net** mass of A **produced** at a point due to chemical reactions per unit volume of material. It is understood that: (i) if species A is produced by chemical reactions, $r_A > 0$, whereas if it is consumed, $r_A < 0$; (ii) r_A contains additive contributions from all the chemical reactions in which species A participates. The two last terms on the right-hand side of Equation 2.32 are quantified as follows:

$$\left\{\begin{matrix}\text{The rate at which mass of species A is created} \\ \text{in the arbitrary volume by chemical reactions}\end{matrix}\right\} - \left\{\begin{matrix}\text{The rate at which mass of species A is consumed} \\ \text{in the arbitrary volume by chemical reactions}\end{matrix}\right\} = \int_{V_a(t)} r_A dV \tag{2.35}$$

Collecting terms, Equation 2.32 becomes

$$\frac{d}{dt}\int_{V_{a(t)}} \rho_A dV = \int_{A_{a(t)}} \rho_A \mathbf{n}\cdot(\mathbf{w}-\mathbf{v}_A)dA + \int_{V_a(t)} r_A dV \tag{2.36}$$

This equation is applicable to any component in the mixture (A = 1, 2, …, N).

At a given point in the material, mass conservation implies that the total mass of species generated by a chemical reaction is compensated by the total mass of species consumed. This imposes the following constraint on the reaction rates

$$\sum_{A=1}^{N} r_A = 0 \tag{2.37}$$

Equation 2.36 also can be expressed in terms of the molar concentration of A, c_A (moles of A per unit mixture volume), by recalling that

$$c_A = \frac{\rho_A}{M_A} \tag{2.38}$$

where M_A is the molecular weight of species A. Dividing Equation 2.36 by M_A yields

$$\frac{d}{dt}\int_{V_{a(t)}} c_A dV = \int_{A_{a(t)}} c_A \mathbf{n}\cdot(\mathbf{w}-\mathbf{v}_A)dA + \int_{V_a(t)} R_A dV \tag{2.39}$$

where

$$R_A = \frac{r_A}{M_A} \tag{2.40}$$

is the molar reaction rate of species A. Note that the constraint imposed by Equation 2.37 can be written as

$$\sum_{A=1}^{N} M_A R_A = 0 \tag{2.41}$$

Equation 2.39 is the **macroscopic mole balance of species** A. The quantity

$$n_A = \int_{V_{a(t)}} c_A dV \tag{2.42}$$

represents the total moles of A in the control volume.

A useful particular case of Equation 2.39 is its application to control volumes that are perfectly mixed. In this case, the reaction rate is uniform throughout the arbitrary volume. Species concentrations are locally uniform over inlet and outlet sections. In addition, good mixing implies that the motion of species at inlet and outlet sections is usually due only to the motion of the mixture, so that $\mathbf{v}_A \approx \mathbf{v}$. Under these conditions, Equation 2.39 can be written as

$$\frac{dn_A}{dt} = \sum_{\text{all inlets i}} Q_i c_{Ai} - \sum_{\text{all outlets i}} Q_i c_{Ai} + V_a R_A \tag{2.43}$$

where

$$Q_i = \left| \int_{A_i} \mathbf{n} \cdot (\mathbf{w} - \mathbf{v}) dA \right| \tag{2.44}$$

is the (always positive) volumetric flow rate that crosses section A_i.

EXAMPLE 2.3 FIRST-ORDER REACTION IN A PERFECTLY MIXED TANK

Consider the case of a compound A that decomposes according to the following chemical reaction:

$$A \rightarrow \text{Products} \tag{2.45}$$

If this reaction is elementary, the disappearance of A follows first-order kinetics and the specific reaction rate is

$$R_A = -kc_A \tag{2.46}$$

An aqueous stream containing $c_{Ain} = 0.1$ mM of a contaminant that decays with a rate given by Equation 2.46 with $k = 1\ h^{-1}$ is fed to a well-stirred holding tank at a rate of 350 L/h. The tank has an overflow that keeps the volume of liquid constant at all times at $V = 50{,}000$ L, and is initially filled with pure water. The tank can be conceptualized as a chemical reactor. This type of system is usually called continuously stirred tank reactor (CSTR).

Since the volume of liquid in the tank is constant, and assuming that changes in concentration do not alter the fluid density, we have: $Q_{in} = Q_{out} = Q = 350$ L/h. A mole balance on A is given by Equation 2.43, which can be expressed as

$$V\frac{dc_A}{dt} = Qc_{Ain} - Qc_A - Vkc_A \tag{2.47}$$

This equation can be separated and integrated from the initial state to an arbitrary point in time

$$V\int_{c_{A0}}^{c_A} \frac{dc_A}{Qc_{Ain} - (Q + Vk)c_A} = \int_0^t dt \tag{2.48}$$

which leads to

$$-\frac{V}{(Q + Vk)}\ln\left(\frac{Qc_{Ain} - (Q + Vk)c_A}{Qc_{Ain} - (Q + Vk)c_{A0}}\right) = t \tag{2.49}$$

Rearranging and noticing that $c_{A0} = 0$ yields

$$c_A = \frac{Qc_{Ain}}{Q + Vk}[1 - e^{-((Q+Vk)/V)t}] \tag{2.50}$$

Substituting known values leads to

$$c_A = 6.542 \times 10^{-3}[1 - e^{-0.0654t}], \quad t \text{ in h and } c_A \text{ in mM} \tag{2.51}$$

It is common to express Equation 2.50 in terms of the **residence time** for the liquid in the tank, defined by

$$\tau = \frac{V}{Q} \tag{2.52}$$

(in this case $\tau = 142.9$ h), which represents the average time that a fluid particle stays in the tank. Equation 2.50 can be rewritten as follows:

$$c_A = \frac{c_{Ain}}{1 + \tau k}[1 - e^{-(1+\tau k)t/\tau}] \tag{2.53}$$

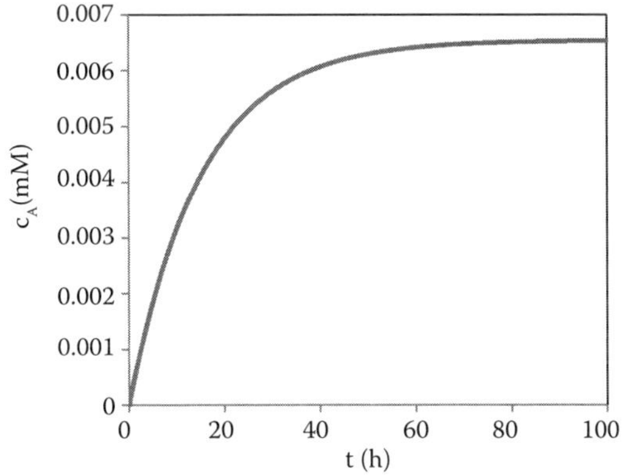

FIGURE 2.6 Evolution of the concentration of species A in the CSTR with chemical reaction (Equation 2.51).

If the system is operated continuously, it will approach a steady state asymptotically. The steady-state concentration can be found by letting $t \to \infty$ in Equation 2.53 to obtain

$$c_{Ass} = \frac{c_{Ain}}{1 + \tau k} = 0.006542\,\text{mM} \tag{2.54}$$

Figure 2.6 shows a plot of the species concentration vs. time (Equation 2.51). The concentration increases monotonically and tends to the steady-state value given by Equation 2.54 asymptotically. Note that, if there were no reaction, we would expect the outlet concentration to reach the inlet concentration (0.1 mM) at steady state. During steady operation, we see that the outlet concentration is 0.6% of the inlet concentration, in which case we say that the reactor is achieving a 99.4% conversion of A.

EXAMPLE 2.4 DECOMPOSITION OF A RADIOACTIVE WASTE

A radioactive waste consisting of an aqueous solution of ^{92}Sr (Strontium 92) at a concentration of 16 mM is pumped at the rate of 1.5 L/min into a perfectly mixed holding tank filled with 400 L of liquid that is initially free of ^{92}Sr. The radioactive decay of ^{92}Sr is a first-order process with $k = 0.257\ h^{-1}$. Liquid flows out of the reactor also at the rate of 1.5 L/min and the liquid density is constant.

1. Plot the molar concentration of ^{92}Sr vs. time. What will be the steady-state concentration?
2. After doing the calculations in 1, you realize that the outlet concentration of ^{92}Sr exceeds a safety limit for ^{92}Sr, which is 5 mM. To avoid exceeding this value, the following operating protocol is suggested: the reactor will be operated with inlet and outlet flows of 1.5 L/min until the outlet concentration reaches 5 mM, at which point the inlet flow will be shut down and the reactor will keep operating with an outflow of 1.5 L/min until it is completely empty. For this mode of operation, find and plot the exit concentration vs. time for the whole operating cycle.

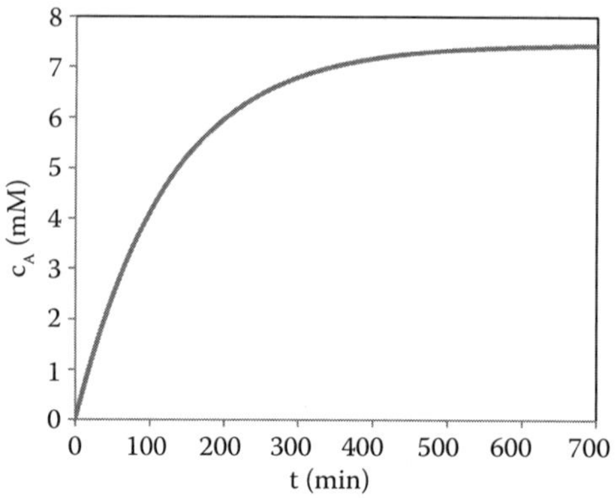

FIGURE 2.7 Evolution of the concentration of ^{92}Sr in the tank for the mode of operation specified in part 1.

1. The first part of this problem is a direct application of Equation 2.50 with $c_{Ain} = 16$ mM, $c_{A0} = 0$, $k = 0.257\ h^{-1} = 4.283 \times 10^{-3}\ min^{-1}$. The result is

$$c_A = 7.47(1 - e^{-8.03\times10^{-3}t}), \quad t \text{ in min}, c_A \text{ in mM} \tag{2.55}$$

   The steady-state concentration ($t \to \infty$) is $c_{Ass} = 7.47$ mM. Figure 2.7 shows how the concentration changes with time in the reactor.

2. The concentration reaches the safety limit $c_A = 5$ mM at $t_s = 137.8$ min, as calculated from Equation 2.55. So Equation 2.55 still applies for $t \leq 137.8$ min. At that point, there is only outflow from the reactor. Hence, from that time on, a total mass balance considering constant density yields

$$\frac{dV}{dt} = -Q \tag{2.56}$$

Since Q is constant, integration of this equation leads to

$$V = V_0 - Qt \tag{2.57}$$

The mole balance in A is

$$\frac{d(Vc_A)}{dt} = -Qc_A - kc_AV \tag{2.58}$$

Expanding the derivative yields

$$V\frac{dc_A}{dt} + c_A\frac{dV}{dt} = -Qc_A - kc_AV \tag{2.59}$$

Substituting Equation 2.57 and simplifying yields

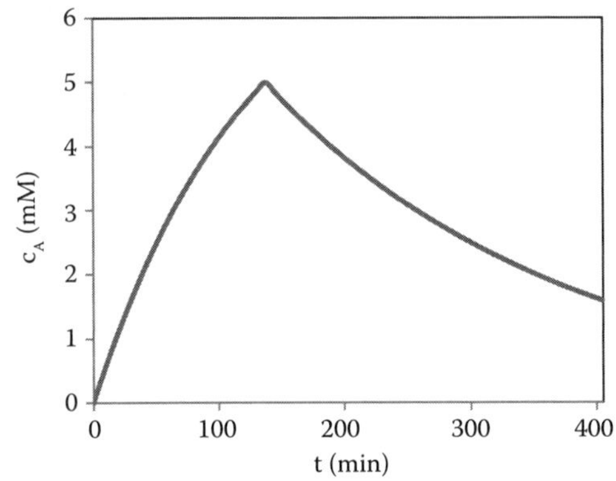

FIGURE 2.8 Evolution of the concentration of ^{92}Sr in the tank for the mode of operation specified in part 2.

$$\frac{dc_A}{dt} = -kc_A \tag{2.60}$$

Note that this is the same equation that would govern a batch reactor (no inlets or outlets). Integrating from an initial condition $t = t_s$, $c_A = c_{As}$, leads to

$$c_A = c_{As}e^{-k(t-t_s)} \tag{2.61}$$

which will apply for $t > t_s = 137.8$ min with $c_{As} = 5$ mM. Note that, at 1.5 L/min, the reactor will be emptied in 400/1.5 = 266.7 min after the flow is shut off, which will be 404.5 min after the start of the process. The whole process can then be represented by

$$c_A = 7.47(1 - e^{-8.03\times10^{-3}t}), \quad t \leq 137.8 \text{ min} \tag{2.62}$$

$$c_A = 5e^{-4.283\times10^{-3}(t-137.8)}, \quad 404.5 \text{ min} > t > 137.8 \text{ min} \tag{2.63}$$

(t in min, c_A in mM). The composite plot of this solution is shown in Figure 2.8.

PROBLEMS

2.1 A CSTR will be used to carry out the chemical decomposition of reactant A, which has a first order rate expression, $R_A = -kc_A$. The constant volume (V) reactor will have inflow and outflow at the same volumetric flow rate, Q. The reactant concentration in the inflow decays exponentially with time, and is given by

$$c_{AI} = \alpha e^{-\beta t}$$

where α and β are constants. Find the concentration in the outflow as a function of time, $c_A(t)$, if there is no A initially in the reactor.

2.2 A tank that has a total volume $V_T = 100$ L and is initially empty will be filled with a liquid flowing at a constant flow rate $Q = 0.1$ L/s. This feed liquid contains a compound A at a concentration $c_{Ai} = 0.1$ mol/L. The compound A decomposes according to a first-order reaction with rate constant $k = 10^{-3}\ s^{-1}$. Find the concentration of A in the tank when it gets completely filled.

Hint: Formulate the mole balance of compound A using as dependent variable the total moles of A in the tank, $n_A(t) = c_A V$.

2.3 The growth of a cell colony in a batch reactor can be simulated by the rate equation

$$\frac{dN}{dt} = \alpha N c_s - \beta N$$

where N is the cell concentration (number of cells per unit volume), c_s is the substrate (nutrient) concentration, and α and β are constants. The first term on the right-hand side of the equation represents the rate of reproductive cell growth while the second term represents a growth inhibition rate due to overpopulation.

The substrate consumption by the cells follows a first-order decomposition reaction, so that

$$\frac{dc_s}{dt} = -kc_s$$

where k is a constant. Initially (t = 0), the cell and substrate concentrations are N_0 and c_{s0}.

a. Find $c_s(t)$ and $N(t)$.
b. A plot of N vs. t reveals that the function has an absolute maximum, which represents the highest possible cell concentration attainable in the culture. Find the time at which this maximum occurs, t_m.
c. Plot N vs. t for the following set of parameters and verify the value of t_m calculated from your solution to part (b):

$N_0 = 30{,}000$ cells/cm^3
$c_{s0} = 0.01$ mol/cm^3
$k = 0.05\ h^{-1}$
$\alpha = 1.5$ cm^3/mol h
$\beta = 0.001\ h^{-1}$

2.4 Consider a perfectly mixed tank containing a constant density liquid in which the following elementary consecutive reactions take place isothermally

$$A \rightarrow B \rightarrow C \qquad R_A = -k_1 c_A;\ R_B = k_1 c_A - k_2 c_B$$

A continuous volumetric flow rate of liquid (Q) is fed into the reactor containing no B or C and a molar concentration of A c_{Ai}. The same flow rate is continuously withdrawn from the reactor, to keep the liquid volume (V) constant. Initially, there is no A, B, or C in the reactor. Find the concentrations of A, B, and C in the reactor as a function of time.

2.5 A solid powder (species A) will be dissolved in a liquid by placing a known mass of solid, m_0, in a liquid volume V that contains initially no A. The solubility of the solid in the liquid is c_{As} (mass concentration) and it is known that the rate of dissolution is given by

$$r = km(c_{As} - c_A)$$

where r represents the mass of solid dissolving per unit time, and m and c_A represent the mass of solid remaining at a given time and the mass concentration of A in the liquid at that time, respectively. Assume that the solution is dilute enough that the volume of liquid remains constant throughout the dissolution process. Find m(t).

2.6 Nitrogen dioxide (NO_2) is a highly toxic gas that is a common by-product of combustion processes. In particular, kerosene heaters that are used indoors are known to liberate NO_2 into the air, which may pose a hazard for people living in the premises. Our objective is to formulate a model for the NO_2 concentration in a house provided with an indoor heater. When it is on, the heater will generate NO_2 at a constant rate S_A (moles per unit time). Ventilation air coming from the outside of the house at a volumetric flow rate Q contains no NO_2, whereas the exit ventilation (also at a rate Q since the volumetric flow rate associated with S_A is negligible) will carry away NO_2 at a concentration equal to that in the house air (see figure). Here, we are assuming that the air in the house is perfectly mixed. Also, nitrogen dioxide decomposes spontaneously in air following a first-order reaction with rate constant k.

a. Find the concentration of NO_2 in the house air as a function of time, $c_A(t)$, if the heater is turned on at $t = 0$ and there is no NO_2 in the house air at that time. What will be the steady-state concentration of NO_2 in the house if the heater is kept continuously on?

b. Now consider that the heater is turned off at a time $t = t_0$. From that point on, ventilation will attenuate the amount of NO_2 in the house air. Determine $c_A(t)$ after the heater is turned off.

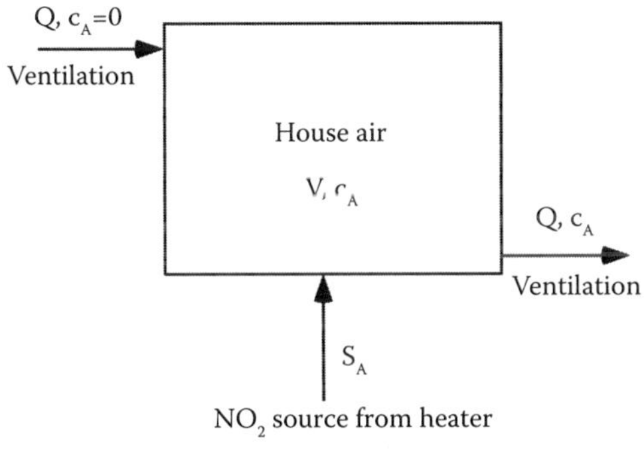

2.7 Radioactive radon gas (Rn^{222}) infiltrates a house from the soil (see figure) at a rate $S_A = 5.8 \times 10^{-18}$ mol/s. The house can be assumed well mixed with a volume $V = 200$ m^3. Ventilation brings air from the outside of the house at a rate $Q = 5$ m^3/h and house air is discharged to the environment at the same rate. The radon decomposes to Polonium-218, producing α particles according to the first-order reaction

$$Rn^{222} \rightarrow Po^{218} + \alpha$$

with $k = 2.1 \times 10^{-6}$ s^{-1}. If there is no radon initially in the house, find the concentration of radon in the house as a function of time.

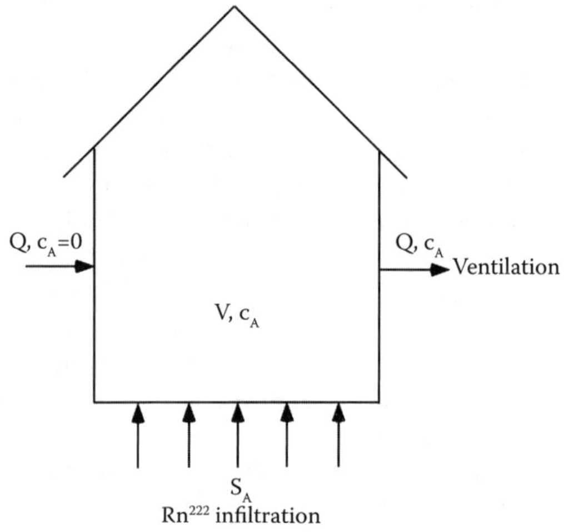

3 Diffusive Transport
Introductory Concepts

3.1 DIFFUSION FUNDAMENTALS

Diffusion is a mass transport mechanism that has profound influence on our daily lives and the world that surrounds us. We often encounter the effects of diffusion with little regard for the technical details of the phenomenon. For example, we are all familiar with the confluence of aromas that emanate from the fragrance counter of a major department store, the dissolution of sugar in a cup of coffee, or the drying of paint on a wall. All of these and many other ordinary, common occurrences are influenced by diffusion. Moreover, the ubiquity of the phenomenon extends to the technical realms of science, medicine, and engineering. Diffusion contributes to the transport of oxygen through living tissues, the presence of sulfur and nitrous oxides in acid rain, the growth of stalactites and related speleothems, the corrosion and degradation of support structures in buildings and bridges, and the time-controlled release of pharmaceuticals in sophisticated drug-delivery schemes. One important feature shared by these various examples is the notion that an individual chemical species can move through a material or medium at a rate or in a direction that differs from that of the medium as a whole. For example, in the absence of circulation or ventilation, the scent from an open bottle of perfume will quickly fill a room. Even if left unstirred, sugar can still dissolve and mix through a cup of coffee. This motion or transport of individual chemical species is generically referred to as diffusion and the phenomenon is an observable consequence of unseen events on the molecular scale.

From an engineering standpoint, diffusion finds application in separation processes, mixing processes, processes that involve phase change, and processes in which chemical reaction occurs. In this chapter, we present a detailed explanation of diffusion as a mass transfer mechanism, including the mathematical representation of diffusion and its relation to the mass conservation equations developed in Chapter 2.

An example of an engineering system in which diffusion plays a crucial role is the separation process known as pervaporation. This process can be used to separate volatile organic contaminants (VOCs) from water. For example, groundwater contaminated with trichloroethylene (TCE) can be remediated by means of pervaporation (EPA, 1998; Das et al., 2006; Kim et al., 2007). The process is based on the fact that the VOC to be separated is relatively volatile (as is TCE at ambient conditions). Pervaporation consists of contacting the contaminated liquid (water + TCE) with a hydrophobic membrane (Figure 3.1). The hydrophobicity of the membrane prevents the water from completely filling the membrane pores, so that a liquid/gas interface exists within the pore space. Due to its volatility, the TCE evaporates into the gas and

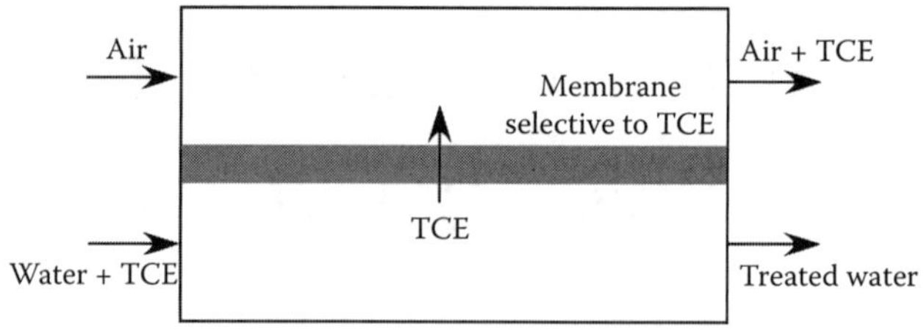

FIGURE 3.1 A cross-flow pervaporation process to separate TCE from water: liquid water contaminated with TCE flows along a membrane selective for TCE; that is, TCE can diffuse through the membrane but other compounds (e.g., less volatile compounds) cannot. The TCE evaporates at a gas/liquid interface in the membrane pores, diffuses across the membrane and is swept away by a stream of air that contacts the membrane.

is transported to the other end of the membrane by diffusion through the gas phase in the pores. The outputs of the process are water depleted of TCE and an air stream with evaporated TCE, which then can be treated, for example, by adsorption of the TCE onto activated carbon. The degree of separation that can be achieved will be determined by the rate at which TCE diffuses through the membrane. This specific example will be analyzed later in this chapter.

Section 3.2 introduces diffusion from a molecular standpoint and presents a framework for the mathematical quantification of diffusive processes. In Section 3.3 we apply these concepts, along with the macroscopic mass balances treated in Chapter 2, to solve case studies that involve one-dimensional problems in which diffusion is the main mechanism of transport.

3.2 DIFFUSION AS A MASS TRANSFER MECHANISM

To expose the physical basis of diffusion, consider the following experiment: a reservoir is divided into two sections by a partition that separates two gases. In one section there is a gas mixture containing molecules of two different chemical species: A and B. In the other section, there is pure B (Figure 3.2). If the partition is removed, random molecular motion will induce molecules of A and B to occupy all the space available. Molecules of A will start appearing in the right side of the reservoir. After a sufficient time, a new equilibrium state is reached (Figure 3.2b) in which there is a uniform molecular distribution (in the average sense) of both A and B over the whole reservoir.

In the realm of the continuum, we are not interested in describing what happens to every molecule, but rather we hope to analyze the processes from a macroscopic point of view. If we look at this process macroscopically (i.e., at the continuum level), we see that there has been a net flow of species A from left to right. This macroscopic transport of A is due to its random molecular motion, and it results in a net transport of A from regions of relatively high concentrations to regions of low concentrations. This transport mechanism is termed **diffusion**.

In the example illustrated by Figure 3.2, a simple mass or mole balance of A can be used to calculate the final concentration of A in the final equilibrium state

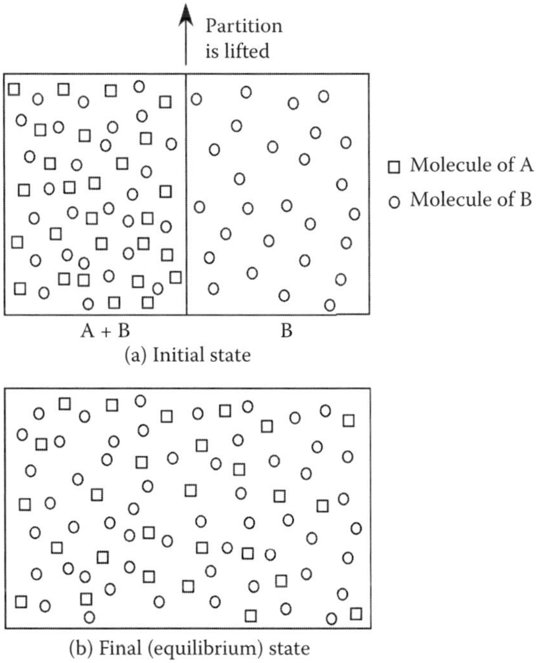

FIGURE 3.2 A solid partition separates two sections of a reservoir, which are each filled with gas. (a) Initially, the section on the left contains molecules of two different chemical species, A and B, while the section on the right contains only molecules of B. The partition is slowly lifted and the system is allowed to evolve toward a new equilibrium state (b), characterized by a uniform distribution of both species throughout the tank. The representation is intended to be schematic; in reality, there will be many more molecules and they will be in a state of continuous motion.

from a knowledge of the initial concentration of A in the left section. However, our interest in the study of mass transfer would be not only to quantify chemical species distribution in equilibrium states, but also to determine the rate at which the process evolves.

Pioneering work on the macroscopic observation of mixing of gases in processes similar to the experiment considered here (Figure 3.2) has been attributed to Graham (1829, 1833, as cited by Cussler, 1997, and Philibert, 2006). Specifically, Graham recognized that gases in this situation would not be spontaneously segregated by their density but rather would mix until a uniform species distribution is achieved.

Diffusion is not limited to gases. There is also spontaneous diffusion of miscible components in liquids and solids, since individual molecules are generally in motion. In the liquid phase, intermolecular interactions decrease the speed at which diffusion takes place, as compared to the gas phase (because of interactions with other molecules, a single molecule does not have the freedom of movement that it would have in a gas). The process is dramatically slower in solids, where molecular motion is generally much more constrained. In a simplified way, it is possible to compare typical times for diffusive transport in gases, liquids, and solids on an order-of-magnitude basis (Table 3.1). Without exposing details of the comparison, diffusion times in liquids are roughly three orders of magnitude longer than in gases, while diffusion times in solids vary widely but can reach several orders of magnitude more than in liquids.

One of the most important characteristics of the diffusion process is that it tends to make the concentration of a chemical species in a mixture uniform. This implies that the transport process is manifested at the macroscopic (continuum) level as a transport from regions of high concentration to regions of low concentration. In other words, the diffusion process can be considered as a macroscopic (continuum level) manifestation of a microscopic (molecular level) process, namely,

Microscopic (molecular-level) aspect: diffusion results from the spontaneous motion of a species through a mixture due to random molecular motion.

Macroscopic (continuum-level) aspect: The diffusing species experiences a net motion from regions of high concentration to regions of low concentration.

Interestingly, the understanding of the macroscopic aspect of diffusion precedes historically the microscopic interpretation. After Graham's observations and measurements of diffusion in gases and liquids, Fick (1855a, b) proposed a mathematical description of macroscopic transport. At the time, it was already known that relatively small particles suspended in a fluid followed seemingly random trajectories, a phenomenon that came to be known as Brownian motion (Brown, 1828). The link between random molecular motion and macroscopic transport by diffusion was formulated mathematically in 1905 by Einstein (as cited by Einstein, 1956) and Smoluchowski (1906).

In what follows, we will explore how to describe quantitatively the diffusion process. Consider the motion of a chemical species (A) in a mixture. In an arbitrarily small volume of mixture, ΔV, there will be N_{mA} molecules of A. Here, ΔV is selected in such a way that it is so small that it looks like a point in the continuum; that is, it is much smaller than a volume representative of the observation scale, but it is big enough that it contains a statistically significant number of molecules of A. The

TABLE 3.1
Order of Magnitude of Diffusion Times in Various Phases

Phase	Diffusion Time through 1 m of Material	Observations
Gas	5 min	Almost no molecular interactions
Liquid	1 day	Molecular interactions are moderate
Solid[a]	Days to months	Molecular interactions are strong

Diffusion times in this table are estimates of the time necessary for an appreciable amount of a chemical species to be transported through 1 m of the phase indicated.

[a] Diffusion times in solids vary greatly with their microscopic structure (e.g., degree of crystallinity).

motion of A through space is quantified by the average velocity vector of all the molecules of A,

$$\mathbf{v}_A = \frac{1}{N_{mA}} \sum_{j=1}^{N_{mA}} \mathbf{v}_{Aj} \tag{3.1}$$

where $\mathbf{v}_{Aj}$ is the velocity of the jth molecule of A present in ΔV. Equation 3.1 is the definition of **species velocity**. Note that, for N_{mA} sufficiently large, $\mathbf{v}_A$ can be considered as a continuum (macroscopic) property. We can use the species velocity to quantify the amount of A that is moving through space. To do this, consider the amount of A that crosses a small surface ΔA perpendicular to the vector $\mathbf{v}_A$ (Figure 3.3). In a time interval Δt, the molecules that cross ΔA are the molecules of A contained in a volume (ΔV) spanned by ΔA and the vector $\Delta\mathbf{r} = \mathbf{v}_A\Delta t$. If p_A is the number concentration of A in the volume (number of molecules of A per unit volume), we can state

$$\left\{\begin{matrix}\text{Molecules of A that cross the} \\ \text{surface } \Delta A \text{ in the period } \Delta t\end{matrix}\right\} = p_A \Delta A \Delta r = p_A \Delta A v_A \Delta t \tag{3.2}$$

where Δr represents the magnitude of the vector **Δr**. If we divide this equation by ΔtΔA (noting that Δt, ΔA → 0), we obtain the molecules of A travelling through space per unit surface area and unit time **at a point in the material**. This is the **molecular flux of species A**,

$$\left\{\begin{matrix}\text{Molecules of A moving through} \\ \text{space per unit time and surface area}\end{matrix}\right\} = p_A \mathbf{v}_A \tag{3.3}$$

It is more convenient to represent species transport in terms of molar concentration (moles of A per unit volume of mixture),

$$c_A = \frac{p_A}{A_v} \tag{3.4}$$

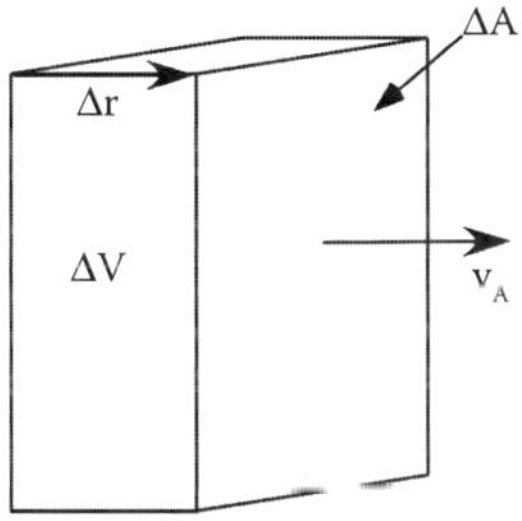

FIGURE 3.3 A small surface in space (ΔA) perpendicular to the species velocity v_A. The volume ΔV is the region of space spanned by the motion of molecules of A in a time Δt.

where A_v is Avogadro's number (6.02×10^{23} molecules/mol). Dividing Equation 3.3 by A_v leads to the definition of **molar flux of A**,

$$\mathbf{N}_A = c_A \mathbf{v}_A \tag{3.5}$$

where $\mathbf{N}_A$ represents the moles of A that move through space per unit time and surface area at a point.

If we consider a surface area perpendicular to the direction of transport of A (i.e., perpendicular to $\mathbf{N}_A$), we can quantify the moles of A that cross the surface per unit time from

$$\mathbf{W}_A = \mathbf{N}_A \Delta A \tag{3.6}$$

where $\mathbf{W}_A$ is the **molar flow rate of A** that crosses the surface ΔA. Note that $\mathbf{N}_A$ is a point (intensive) property, since we can assign a value for it at each point in space, whereas $\mathbf{W}_A$ is a macroscopic (extensive) property, since its value depends on the size of the surface ΔA.

For now, we will deal with a binary mixture made up of two species, A and B, in which the motion of B will be quantified by the molar flux of B, $\mathbf{N}_B$, defined analogously to the molar flux of A. The total molar flux (net moles of mixture transported per unit time and per unit area) is the sum of the fluxes of the species comprising the mixture,

$$\mathbf{N} = \mathbf{N}_A + \mathbf{N}_B \tag{3.7}$$

where

$$\mathbf{N}_B = c_B \mathbf{v}_B \tag{3.8}$$

An average velocity for the mixture, $\mathbf{v}^*$, can be defined using an equation similar to Equations 3.5 and 3.8,

$$\mathbf{N} = \mathbf{v}^* c \tag{3.9}$$

On the other hand, the total molar concentration (or molar density) of the mixture is the sum of the concentrations of the two species,

$$c = c_A + c_B \tag{3.10}$$

Combining Equations 3.5, 3.7, 3.8, and 3.9 leads to

$$\mathbf{v}^* c = \mathbf{v}_A c_A + \mathbf{v}_B c_B \tag{3.11}$$

The mole fractions of the two species are given by

$$x_A = \frac{c_A}{c}, \quad x_B = \frac{c_B}{c} \tag{3.12}$$

Using these relations, Equation 3.11 can be written as

$$\mathbf{v}^* = x_A \mathbf{v}_A + x_B \mathbf{v}_B \tag{3.13}$$

so that the mixture velocity defined by its use in Equation 3.9, $\mathbf{v}^*$, is the **molar averaged velocity** of the mixture.

When the principles of mass and momentum conservation are used in the study of Fluid Mechanics (see Chapter 6), motion is represented by a velocity indicative of mass flow, instead of molar flow. Here, we will generalize the concept to mixtures. Let $\mathbf{n}_A$ and $\mathbf{n}_B$ be the **species mass fluxes** (mass of species transported per unit time and surface area at a point in space). We can obtain these fluxes from

$$\mathbf{n}_A = M_A \mathbf{N}_A \tag{3.14}$$

$$\mathbf{n}_B = M_B \mathbf{N}_B \tag{3.15}$$

where M_A and M_B are the species molecular weights. We can also define a total mass flux of the mixture by

$$\mathbf{n} = \mathbf{n}_A + \mathbf{n}_B \tag{3.16}$$

Multiplying Equations 3.5 and 3.8 by the species molecular weights and using Equations 3.14 and 3.15, we find

$$\mathbf{n}_A = c_A M_A \mathbf{v}_A \tag{3.17}$$

$$\mathbf{n}_B = c_B M_B \mathbf{v}_B \tag{3.18}$$

The product between the molecular weight and the molar concentration is the mass concentration (mass of species per unit volume of mixture),

$$\rho_A = c_A M_A \tag{3.19}$$

$$\rho_B = c_B M_B \tag{3.20}$$

So that Equations 3.17 and 3.18 can be written as

$$\mathbf{n}_A = \rho_A \mathbf{v}_A \tag{3.21}$$

$$\mathbf{n}_B = \rho_B \mathbf{v}_B \tag{3.22}$$

An average velocity for the mixture, **v**, can be defined using an equation similar to these two equations,

$$n = \rho v \tag{3.23}$$

The total mass concentration (or density) is the sum of the mass concentrations of the two species,

$$\rho = \rho_A + \rho_B \tag{3.24}$$

Combining Equations 3.16, 3.21, 3.22, and 3.23 leads to

$$\mathbf{v}\rho = \mathbf{v}_A\rho_A + \mathbf{v}_B\rho_B \tag{3.25}$$

The mass fractions of the two species are given by

$$w_A = \frac{\rho_A}{\rho}, \quad w_B = \frac{\rho_B}{\rho} \tag{3.26}$$

Using these relations, Equation 3.25 can be written as

$$\mathbf{v} = w_A\mathbf{v}_A + w_B\mathbf{v}_B \tag{3.27}$$

This is the **mass-averaged velocity** of the mixture. Note that, even though both velocities (molar and mass-averaged velocities) represent the motion of the mixture, in general $\mathbf{v}^* \neq \mathbf{v}$. There are two noteworthy cases in which the two velocities are the same: (i) when $M_A = M_B$ (e.g., if A and B are isomers); (ii) when one of the components is dilute (e.g., if $x_A \ll 1$, the mixture is almost pure B and the previous equations can be used to show that $\mathbf{v}^* \approx \mathbf{v} \approx \mathbf{v}_B$).

In the study of mass transfer, the molar representation of fluxes and concentrations usually is preferred over the mass representation, especially when chemical reactions are involved, since information on chemical kinetics and chemical equilibrium usually is expressed on a molar basis.

The preceding definitions and equations are generalized directly to a multicomponent system (mixture of N_c components) as follows:

$$\mathbf{N}_A = \mathbf{v}_A c_A, \quad A = 1,2\ldots,N_c \tag{3.28}$$

$$\mathbf{N} = \sum_{A=1}^{N_c} \mathbf{N}_A = c\mathbf{v}^* \tag{3.29}$$

$$\mathbf{v}^* = \sum_{A=1}^{N_c} x_A \mathbf{v}_A \tag{3.30}$$

$$\mathbf{n}_A = \mathbf{v}_A \rho_A, \quad A = 1,2,\ldots,N_c \tag{3.31}$$

$$\mathbf{n} = \sum_{A=1}^{N_c} \mathbf{n}_A = \rho \mathbf{v} \tag{3.32}$$

$$\mathbf{v} = \sum_{A=1}^{N_c} w_A \mathbf{v}_A \tag{3.33}$$

The motion of species A in the mixture is a consequence of two distinct mechanisms: diffusion and convection. As mentioned before, diffusion is the transport of the chemical species due to random molecular motion. Convection refers to the motion of the chemical species because of bulk motion of the mixture. If we represent the motion using a molar basis, convection will be quantified by means of the mixture velocity $\mathbf{v}^*$. Since the two mechanisms are independent, we can decompose the species velocity as follows:

$$\mathbf{v}_A = \mathbf{v}^* + \mathbf{u}_A^* \tag{3.34}$$

This equation defines $\mathbf{u}_A^*$ as the **molar diffusion velocity** of species A. Note that the diffusion velocity is the velocity of A **relative to** the velocity of the mixture; that is, diffusive motion is superimposed to the motion of the mixture. Using $\mathbf{v}_A$ from this equation into Equation 3.5 leads to

$$\mathbf{N}_A = c_A \mathbf{v}^* + c_A \mathbf{u}_A^* \tag{3.35}$$

The first term on the right-hand side of this equation is the **molar convective flux** of species A, and the second term is the **molar diffusive flux**, which we will denote by

$$\mathbf{J}_A^* = c_A \mathbf{u}_A^* \tag{3.36}$$

Using Equation 3.9, the convective flux in Equation 3.35 can be expressed in terms of the total molar flux (**N**), so that Equation 3.35 becomes

$$\mathbf{N}_A = x_A \mathbf{N} + \mathbf{J}_A^* \tag{3.37}$$

In a binary system, a similar equation can be found for component B:

$$\mathbf{N}_B = x_B \mathbf{N} + \mathbf{J}_B^* \tag{3.38}$$

where

$$\mathbf{J}_B^* = c_B \mathbf{u}_B^* \tag{3.39}$$

and the molar diffusion velocity of species B is

$$\mathbf{u}_B^* = \mathbf{v}_B - \mathbf{v}^* \tag{3.40}$$

For this binary mixture we see that, if we add Equations 3.37 and 3.38, the left-hand side will become the total molar flux (Equation 3.7) and

$$\mathbf{N} = (x_A + x_B)\mathbf{N} + \mathbf{J}_A^* + \mathbf{J}_B^* \tag{3.41}$$

Since mole fractions must add up to 1, we conclude

$$\mathbf{J}_A^* = -\mathbf{J}_B^* \tag{3.42}$$

that is, the molar diffusive fluxes of A and B have the same magnitude but opposite direction. Another way to say this is that the **net** molar diffusive flux of the mixture is zero. This conclusion is a consequence of the fact that diffusive motion is superimposed on the motion of the mixture. In other words, if we were to use a coordinate system that moves with the mixture, all motion would be due to diffusion and the mixture as a whole would be stationary.

For a multicomponent system, the equivalent version of Equation 3.42 can be shown to be

$$\sum_{A=1}^{N_c} \mathbf{J}_A^* = 0 \tag{3.43}$$

Molar diffusive fluxes have the following characteristics:

1. They are a consequence of processes that happen at the molecular level (diffusion).
2. The direction of the diffusive flux vector is from regions of high concentration to regions of low concentration.

The first characteristic implies that a mathematical representation for the diffusive fluxes must be based on a **constitutive equation** that reflects random molecular motion. The second characteristic indicates that the flux vector will have the opposite direction of the concentration gradient of the species. In the establishment of the constitutive equation, we will consider three different cases.

3.2.1 Binary Systems

For systems with two components (A and B), the constitutive equation for the diffusive flux is **Fick's law**,

$$\mathbf{J}_A^* = -cD_{AB}\nabla x_A \tag{3.44}$$

According to this equation, the diffusive flux is parallel and directly proportional to the mole fraction gradient. The minus sign reflects the fact that transport will occur from high-to-low concentrations (see discussion of Figure 3.4). The coefficient D_{AB} is the **diffusivity or binary diffusion coefficient** of A in B. A similar expression applies to component B,

$$\mathbf{J}^*_B = -cD_{BA}\nabla x_B \tag{3.45}$$

where D_{BA} is the diffusivity of B in A. Note that, since

$$x_A + x_B = 1 \Rightarrow \nabla x_A + \nabla x_B = 0 \tag{3.46}$$

Addition of Equations 3.44 and 3.45 using Equations 3.42 and 3.46 leads to

$$0 = -c(D_{AB} - D_{BA})\nabla x_A \tag{3.47}$$

Since this equation must be valid for any mole fraction field, we conclude that

$$D_{AB} = D_{BA} \tag{3.48}$$

that is, one diffusion coefficient is enough to quantify diffusion in a binary mixture.

The total molar concentration or molar density, c, can be assumed to be uniform for isothermal liquids and for gases with uniform temperature and pressure (recall that, for an ideal gas, c = P/RT). If this is the case, c can be placed into the gradient in Equation 3.44 to get

$$\mathbf{J}^*_A = -D_{AB}\nabla c_A, \quad \text{uniform c} \tag{3.49}$$

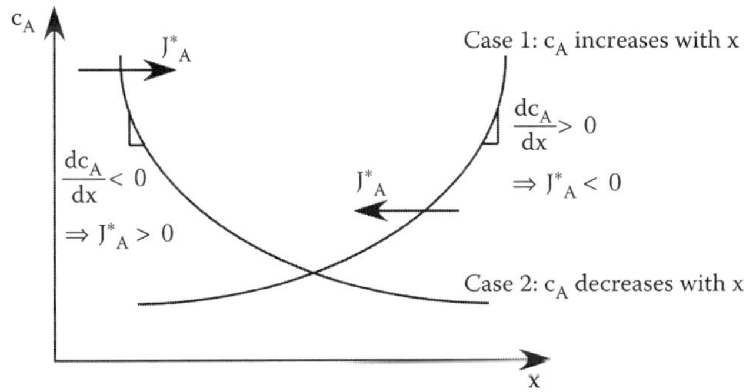

FIGURE 3.4 The diffusive flux in relation to the concentration profile for transport in one direction (x). The only nonzero component of the concentration gradient in Equation 3.49 is, in this case, dc_A/dx, which leads to a flux in the x direction: $J^*_A = -D_{AB}dc_A/dx$. The sign of the flux reflects the direction in which A diffuses: Case 1, c_A increases with x so that the concentration gradient is positive and the flux is negative (A diffuses in the negative x-direction); Case 2, c_A decreases with x so that the concentration gradient is negative and the flux is positive (A diffuses in the positive x-direction).

As noted before, the minus sign dictates that the direction of the flux is from high-to-low concentrations. Figure 3.4 illustrates this for the case of one-dimensional transport in the x direction.

The diffusivity, D_{AB}, represents the link between the random molecular motions that cause diffusion and the diffusive flux, which is their macroscopic manifestation. In general, the diffusivity depends on

1. The chemical structures of species A and B
2. The thermodynamic state of the system (e.g., T and P)

Note that, the higher the diffusivity, the higher the flux for a given concentration gradient. Examples of diffusivities of binary systems are given in Table 3.2.

An important characteristic of the diffusivity is that its magnitude depends on the phase in which the diffusive transport occurs. Diffusivities in gases are usually of the order of 0.1 cm²/s, while in liquids they are of the order of 10^{-5} cm²/s. This difference reflects a stronger level of intermolecular interactions in the liquid phase, which makes diffusion slower (recall the comparison of diffusion times in Table 3.1). In gases, the diffusivity decreases as the size of the diffusing molecules increases. For example, the binary diffusivity of the H_2/CH_4 system is significantly higher than that of the CO_2/N_2 system. Binary gas mixtures involving H_2 exhibit relatively high diffusivities because the dihydrogen molecule is amongst the smallest.

For ideal gases at relatively low temperatures (e.g., Bird et al. 2007), the kinetic theory of gases yields the prediction

$$D_{AB} \sim \frac{T^2}{P}\sqrt{\frac{1}{M_A}+\frac{1}{M_B}} \tag{3.50}$$

where the effect of molecular size is evident (as reflected by the molecular weights). Higher temperatures lead to more rapid molecular motion, more energetic collisions

TABLE 3.2
Binary Diffusivities at 298 K and 1 atm

Components (Phase)	D_{AB} (cm²/s)
H_2/CH_4 (gas)	0.726
CO_2/N_2 (gas)	0.167
O_2/H_2O (liquid)[a]	2.41×10^{-5}
KCl/H_2O (liquid)[a]	1.87×10^{-5}
KCl/ethylene glycol (liquid)[a]	0.12×10^{-5}
CO_2/rubber (solid)[a]	1.1×10^{-6}
He/Fe (solid)[a]	2.6×10^{-9}
Al/Cu (solid)	1.3×10^{-30}

[a] The solute, listed as the first component, is dilute.

and, consequently, faster diffusion. Note also that an increase in pressure leads to a lower diffusivity. Higher pressures increase the total molar concentration of the mixture, which diminishes the mean-free path between collisions and increases the probability of intermolecular collisions. Accordingly, for a given time interval, the root-mean displacement of individual molecules decreases and the diffusivity consequently decreases.

Molecular theories of diffusion in liquids usually consider the diffusing molecules as particles moving through a liquid continuum. Molecular motion is affected by the friction of the diffusing molecule with the liquid, which establishes a link between diffusivity and viscosity. An empirical equation to estimate the diffusivity of A in B in a solution in which A is dilute ($x_A \ll 1$) is the Wilke–Chang equation, namely,

$$D_{AB} \approx 7.4 \times 10^{-8} T \frac{\sqrt{\psi_B M_B}}{\mu \tilde{V}_A^{0.6}} \quad (3.51)$$

where μ is the solution viscosity (which would be approximately the viscosity of liquid B) in mPa s, $\tilde{V}_A$ is the molar volume of A (cm^3/mol) as a liquid at its normal boiling point, ψ_B is an association parameter for the solvent, M_B is the molecular weight of the solvent B in g/mol, T is in K, and D_{AB} is in cm^2/s. Recommended values for ψ_B include (Bird et al., 2007): water, 2.6; methanol, 1.9; and benzene and heptane, 1.0.

The inverse relation between diffusivity and viscosity is reflected by the two liquid-phase diffusivities quoted in Table 3.2 for KCl; at these conditions the viscosity of water is 0.9 mPa s, while the viscosity of ethylene glycol is about 15 mPa s. Other methods and relationships to predict diffusivities in gases and liquids are reviewed by Reid et al. (1987).

Diffusivities in solids are strongly influenced by the molecular order of the solid phase, and their order of magnitude can vary widely, as seen in the examples shown in Table 3.2. In general, soft amorphous solids, such as rubbery polymers, tend to behave similarly to liquids, with diffusivities close to the order of magnitude expected for liquids (e.g., CO_2/rubber), while diffusion of relatively large atoms through crystalline solids have diffusivities that can be many orders of magnitude lower (such as Al/Cu).

3.2.2 Pseudo-Binary Systems

As we will see below, constitutive equations for multicomponent systems are considerably more complex than Fick's law. However, the transport of a species in a multicomponent system sometimes can be approximated by Fick's law. Such systems will be called pseudo-binary systems. We will distinguish two cases:

1. Systems in which the transported species of interest is dilute ($x_A \ll 1$).
2. Systems in which the rest of the components in the mixture (except A) keep their relative proportions. An example of this is the transport of a gas through air. As long as the relative proportions of oxygen and nitrogen in the air are the same, the mixture can be assumed to behave as a single component.

In both cases, Fick's law applies:

$$\mathbf{J}_A^* = -D_{Am}\nabla c_A, \text{uniform c} \tag{3.52}$$

and the diffusion coefficient, D_{Am}, is the diffusivity of A in the mixture.

3.2.3 Multicomponent Systems

To understand the mathematical form of the constitutive equation for multicomponent systems, consider again Fick's law for a binary system. From Equations 3.36 and 3.44, we can get an expression for the molar diffusion velocity of A,

$$\mathbf{u}_A^* = -\frac{1}{x_A}D_{AB}\nabla x_A \tag{3.53}$$

Similarly, for component B (recalling Equation 3.48)

$$\mathbf{u}_B^* = -\frac{1}{x_B}D_{AB}\nabla x_B \tag{3.54}$$

Subtracting Equations 3.53 and 3.54 and rearranging leads to

$$\frac{x_A x_B}{D_{AB}}(\mathbf{u}_B^* - \mathbf{u}_A^*) = -x_A\nabla x_B + x_B\nabla x_A \tag{3.55}$$

Using Equation 3.46, we see that

$$-x_A\nabla x_B + x_B\nabla x_A = x_A\nabla x_A + (1 - x_A)\nabla x_A = \nabla x_A \tag{3.56}$$

so that Equation 3.54 leads to

$$\nabla x_A = \frac{x_A x_B}{D_{AB}}(\mathbf{u}_B^* - \mathbf{u}_A^*) \tag{3.57}$$

We emphasize that this equation is a consequence of Fick's law for a binary system. For multicomponent systems, there are no constitutive equations that are explicit in the flux of each species. The constitutive equations have a mathematical structure that is consistent with Equation 3.57; namely,

$$\nabla x_B = \sum_{A=1}^{N_c} \frac{x_B x_A}{D_{BA}^*}(\mathbf{u}_A^* - \mathbf{u}_B^*), \quad B = 1, 2,\ldots,N_c \tag{3.58}$$

where the coefficients D_{BA}^* are not, in general, the binary diffusivities but properties that are associated with the specific mixture. The set of Equations 3.58 for all components are called the **Stefan–Maxwell equations**. They form a set of linear algebraic equations whose solution yields the diffusion velocities for each species, from which the respective diffusive fluxes can be calculated. The Stefan–Maxwell equations are rarely used in the modeling of mass transfer in environmental systems. As we will see, the case studies and practical examples to be treated here fall within the categories of binary or pseudo-binary systems.

3.3 ONE-DIMENSIONAL, STEADY DIFFUSION

In this section, we solve a series of problems to illustrate the use of the fundamental relationships developed in the previous sections. We restrict our discussion to systems in which the salient diffusive processes are unidirectional, that is, the diffusive flux vector has a single nonzero component. We will consider that diffusion occurs only in the x-direction and that the mass transfer process is at steady state (i.e., intensive properties do not change with time).

We begin in a rather general way by considering the one-dimensional transport of a chemical species A through a thin control volume V_a, as depicted in Figure 3.5. According to Equation 2.39, a mole balance of A applied to the control volume reads

$$\frac{d}{dt}\int_{V_{a(t)}} c_A\, dV = \int_{A_{a(t)}} c_A \mathbf{n}\cdot(\mathbf{w}-\mathbf{v}_A)dA + \int_{V_a(t)} R_A\, dV \qquad (3.59)$$

Now we stipulate that V_a is fixed in space, so that $\mathbf{w} = \mathbf{0}$ everywhere on the control surface, and the system is at steady state so that the left-hand side of Equation 3.59 vanishes. We also have

$$\mathbf{n}\cdot\mathbf{v}_A = -v_A\big|_x, \quad \text{at x} \qquad (3.60)$$

$$\mathbf{n}\cdot\mathbf{v}_A = v_A\big|_{x+\Delta x}, \quad \text{at x} + \Delta x \qquad (3.61)$$

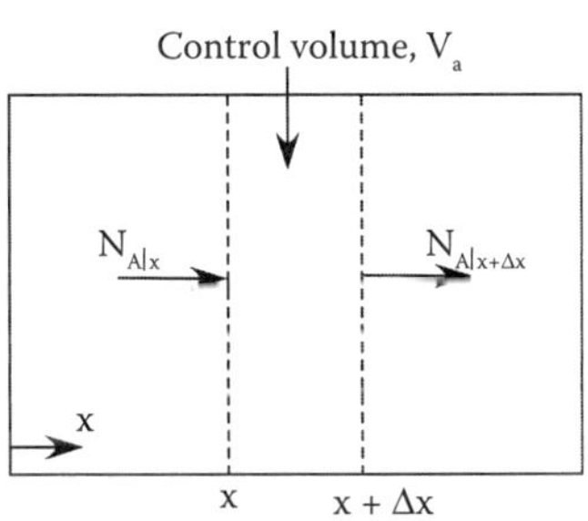

FIGURE 3.5 Definition sketch for the mole balance on species A for one-dimensional transport in the x direction. The control volume V_a extends an infinitesimal distance Δx in the direction of mass transport.

where we use v_A to represent the x-component of the species velocity, which is the only nonzero component of the vector. The geometry of the control volume is such that $V_a = S\Delta x$, where S is the cross-sectional area, and we assume that c_A and $\mathbf{v}_A$ depend on x, but do not vary in directions orthogonal to x. Subject to these considerations, Equation 3.59 reduces to

$$0 = c_A v_A \big|_x S - c_A v_A \big|_{x+\Delta x} S + R_A S\Delta x \tag{3.62}$$

where R_A has been assumed to be uniform in the small control volume. The three terms of this equation can be interpreted in terms of a mole balance of species A:

$$\begin{aligned} 0 = &\begin{Bmatrix} \text{Moles of A entering the} \\ \text{control volume per unit time} \end{Bmatrix} - \begin{Bmatrix} \text{Moles of A leaving the} \\ \text{control volume per unit time} \end{Bmatrix} \\ &+ \begin{Bmatrix} \text{Net number of moles of A generated} \\ \text{in the control volume by chemical reaction} \end{Bmatrix} \end{aligned} \tag{3.63}$$

Dividing Equation 3.62 by $S\Delta x$ and taking the limit as $\Delta x \to 0$ yields

$$\frac{d(c_A v_A)}{dx} = R_A \tag{3.64}$$

Equivalently, using Equation 3.5 and letting N_A be the x-component of the molar flux of A, we can state

$$\frac{dN_A}{dx} = R_A \tag{3.65}$$

Note that this equation is also applicable to species B in a binary mixture,

$$\frac{dN_B}{dx} = R_B \tag{3.66}$$

An important implication of Equations 3.65 and 3.66 is that, in the absence of chemical reactions (i.e., $R_A = R_B = 0$), N_A and N_B are independent of x; that is, mass conservation states that the (steady-state) flux of species will not change in the direction of motion if the species is not generated or consumed by chemical reactions.

The species molar flux can be related to species concentration by combining Equation 3.37 with Fick's law. For a binary system and transport in the x direction only, use of Equation 3.44 in Equation 3.37 yields

$$N_A = x_A(N_A + N_B) - cD_{AB}\frac{dx_A}{dx} \tag{3.67}$$

with a similar equation applicable to species B.

In what follows, we consider specific applications dealing with one-dimensional, steady diffusion.

3.3.1 Diffusion through a Solid Membrane

Consider the mass transfer process that occurs when a chemical species A diffuses through a solid membrane B as depicted in Figure 3.6. The membrane is interposed between two well-mixed gases of different compositions. Here, we represent the content of A in the gases using its partial pressure, P_A, and we consider that $P_{A1} > P_{A2}$. In our analysis, we assume that the membrane is perfectly selective with respect to A so that no other species in the gas can pass through the membrane. We also assume that the partial pressures P_{A1} and P_{A2} are both steady and known.

Our immediate goal is to develop a quantitative description of the diffusive transport of A through the membrane. Before doing this, however, it is worth emphasizing that, as posed, the problem contains an important feature common to a large number of mass transfer applications, namely, the transport of a chemical species across a phase boundary (in this case the gas/membrane interfaces at x = 0 and L). In terms of the coordinate system laid out Figure 3.6, species A must pass from a gas phase into the membrane at x = 0, conversely, molecules of A must pass from the solid to a gas at x = L. To analyze and understand mass transport processes, one typically wants or needs to know the concentration distribution of the species in flux. Here, we will want to work out $c_A(x)$, the concentration profile of A in the membrane.

At a phase boundary, such as a gas/solid interface, the concentration of a chemical species is generally not a continuous function of position. Along an axis passing perpendicular to a phase interface, chemical concentration profiles generally exhibit a precipitous jump, or discontinuity. The reasons for the discontinuity in concentrations are multiple and have mainly to do with the differences in the detailed molecular structure and molecular interactions prevalent in the separate continuous phases. These microstructural differences result, on the continuum-level, in a mathematical discontinuity of the species concentration from one side of a phase boundary to the

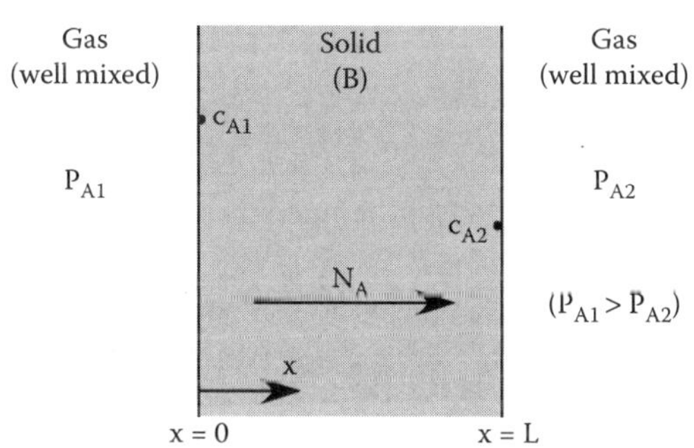

FIGURE 3.6 A solid membrane made of component B separates two gases with different partial pressures of species A, which is soluble in the membrane material. Because of the partial pressure difference, species A will diffuse across the membrane with molar flux N_A, as indicated. Other compounds might be present in the gases on each side of the membrane.

other. Indeed, from the continuum perspective, a phase boundary or interface is an infinitesimally thin, two-dimensional surface at which a number of material properties (e.g., density, diffusivity, electrical, and thermal conductivity) are discontinuous. One practical implication of these discontinuities is evident in the problem at hand. Though we may know or be able to measure the concentration of A in the gas phases of Figure 3.6, we do not necessarily know or have direct experimental measurements of c_{A1} and c_{A2}, the concentrations of A in the membrane at x = 0 and L, respectively. Molecules of A are subject to weak molecular interactions in the gas phase, whereas in the solid phase, they must fit into a more compact arrangement of molecules, where direct interactions with molecules of B are likely. The extent to which the solid phase can accommodate species A, in terms of moles of A per unit volume of material, is generally expected to be far less than that of the gas. The solubility of A in the solid will depend on the compatibility of A and B, as well as the microstructure of the membrane.

We know from thermodynamics that, if neighboring phases are in equilibrium, one can relate component concentrations on one side of the phase boundary to those on the other. To accommodate the jump in species concentration at an interface, then, we introduce the notion of **local equilibrium**. The underlying rationale for the idea is that the first few layers of molecules that sandwich the interface equilibrate rapidly compared to the rates of mass transport in the bulk phases. So, while the entirety of the two phases is decidedly not in equilibrium, it is often assumed that equilibrium holds locally at a phase interface.

Turning to the scenario in Figure 3.6, we invoke the local equilibrium hypothesis and assert that c_{A1} and c_{A2} are known thermodynamic functions of P_{A1} and P_{A2}, respectively. Specifically, we assume the concentration of A in the faces of the membrane corresponds to the solubility of A in the solid at the gas-phase conditions in question; namely, c_{A1} is the solubility of A in B when the solid is exposed to a partial pressure P_{A1}, and c_{A2} is the solubility of A in B when the solid is exposed to P_{A2}. For simplicity, we will consider that the equilibrium (solubility) relation is linear; that is,

$$c_{A1} = KP_{A1} \tag{3.68}$$

$$c_{A2} = KP_{A2} \tag{3.69}$$

where K is a constant that depends on temperature and total pressure, and can be measured independently.

Given Equations 3.68 and 3.69, and knowing that $P_{A1} > P_{A2}$, we infer that $c_{A1} > c_{A2}$ and the flux of A through the membrane will be in the (positive) x direction. In this case, we expect A to be present at relatively low concentrations in the solid so that we can consider the molar density, c, to be uniform in the solid. This allows us to express Equation 3.67 as

$$N_A = x_A(N_A + N_B) - D_{AB}\frac{dc_A}{dx} \tag{3.70}$$

In this case we can set $N_B = 0$ since the solid is stationary, and rearrange this equation to the form

$$(1 - x_A)N_A = -D_{AB}\frac{dc_A}{dx} \tag{3.71}$$

The condition that A is present at low concentrations in the solid (dilute system) can be expressed mathematically by stating that the mole fraction of A everywhere is very small compared to 1; that is, $x_A \ll 1$, which implies: $1 - x_A \approx 1$, which in turn yields

$$N_A = -D_{AB}\frac{dc_A}{dx}, \quad 0 \le x \le L \tag{3.72}$$

Comparing Equations 3.70 and 3.72 we see that, as a result of the dilute approximation, the convection term, $x_A (N_A + N_B)$, has vanished from the problem formulation. Convection plays a negligible role in the transport of A through the membrane because: (i) component B is not in flux, and (ii) A is only present as a dilute constituent in the solid.

If A does not participate in chemical reactions within the membrane, we can set $R_A = 0$ and the species mole balance (Equation 3.65) simplifies to

$$\frac{dN_A}{dx} = 0 \Rightarrow N_A \text{ is independent of } x \text{ for } 0 \le x \le L \tag{3.73}$$

Since N_A is constant, we can separate variables and integrate Equation 3.72 across the membrane, as follows:

$$N_A\int_0^L dx = -D_{AB}\int_{c_{A1}}^{c_{A2}} dc_A \tag{3.74}$$

Evaluating the integrals yields

$$N_A = D_{AB}\frac{(c_{A1} - c_{A2})}{L} \tag{3.75}$$

This equation can be expressed in terms of partial pressures by using Equations 3.68 and 3.69,

$$N_A = KD_{AB}\frac{(P_{A1} - P_{A2})}{L} \tag{3.76}$$

We see from Equations 3.75 and 3.76 that the flux of A through the membrane is directly proportional to the concentration difference $(c_{A1} - c_{A2})$ or the partial pressure difference ($P_{A1} - P_{A2}$). Either of these differences is the **driving force** behind the diffusion process.

Since the partial pressures of A in the gases are known, we can calculate the flux from Equation 3.76. Now we can return to Equation 3.72 and derive an expression for $c_A(x)$. To do this, we integrate the equation from x = 0 to a generic point inside the membrane with coordinate x,

$$N_A \int_0^x dx = -D_{AB} \int_{c_{A1}}^{c_A(x)} dc_A \quad (3.77)$$

Evaluation of the integrals yields

$$N_A x = -D_{AB}(c_A - c_{A1}) \quad (3.78)$$

Substitution of Equation 3.75 and rearranging leads to

$$c_A = c_{A1} - (c_{A1} - c_{A2})\frac{x}{L} \quad (3.79)$$

That is, the concentration of A varies linearly across the membrane. A sketch of this equation is shown in Figure 3.7.

The results obtained here can be generalized to any diffusion process through a flat film or membrane, as long as the diffusing component is present in the under dilute conditions. That is, Equations 3.75 and 3.79 will apply regardless of what phase(s) are in contact with the boundaries at x = 0 and x = L. The effects of the surrounding phases come into play when the concentrations c_{A1} and c_{A2} are related to external concentrations by means of equilibrium relations. For the particular case where the film or membrane is enveloped by gases and the equilibrium relations are linear (Equations 3.68 and 3.69), the molar flux of A can be calculated from Equation 3.76, which is usually expressed as

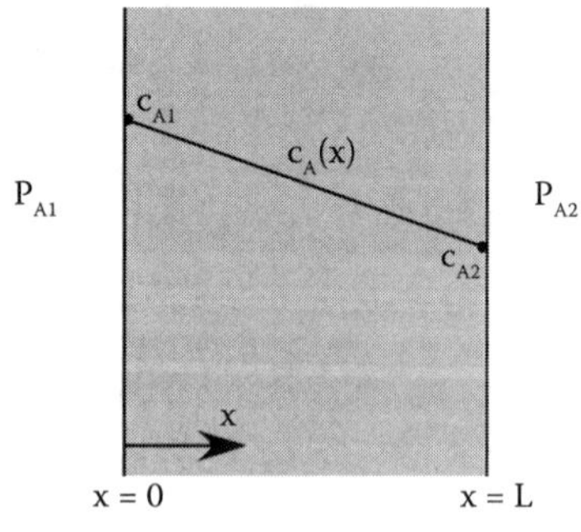

FIGURE 3.7 Concentration profile of species A in the membrane.

$$N_A = K_P \frac{(P_{A1} - P_{A2})}{L} \tag{3.80}$$

where

$$K_P = KD_{AB} \tag{3.81}$$

is called the permeability coefficient.

3.3.2 Analysis of a Pervaporation Process

Membrane processes are widely used in environmental systems. Their main application is in separation processes. In the previous application, we considered a membrane as a continuous phase in which the diffusing component could dissolve. Often, membranes contain pores of a specified size, designed to allow for the passage of a liquid or gas, while precluding or retarding the passage of solid particles or assembled molecular structures, such as colloids, present in the fluid. Particles and molecular assemblages are thereby separated from the fluid that crosses the membrane; in effect, the membrane is a porous medium that functions as a sieve or filter. Fluid motion through the interstices of the membrane is largely hydrodynamic, a flow driven by the imposition of a pressure gradient across the membrane.

An example of a membrane process used in environmental systems is the pervaporation process, described briefly at the beginning of this chapter. A typical pervaporation process is depicted in Figure 3.1, where a liquid consisting of two components, A and D (water and TCE in the example, respectively), is brought into contact with a membrane located between the liquid and an inert carrier gas C (air). Species A and D evaporate at the gas/liquid interface formed in the membrane pores and then diffuse toward the carrier gas (Figure 3.8). If the rates at which A and D are transferred to the gas are different, there will be an enrichment of one component relative to the other in the liquid phase; the liquid will become richer in the component whose transfer rate is slower. To analyze the pervaporation process, we will suppose that species A and D are present in dilute quantities in the gas phase

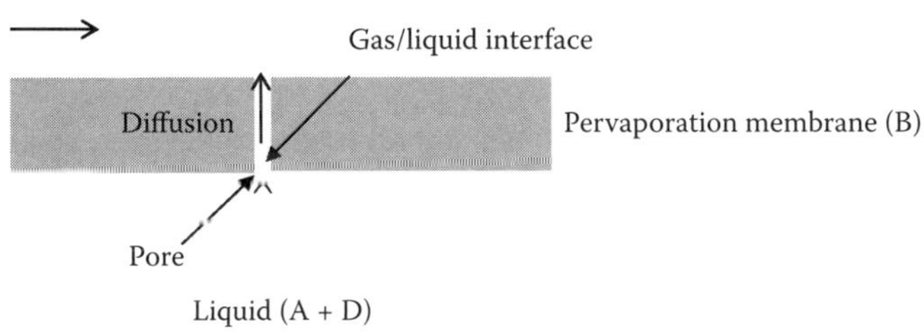

FIGURE 3.8 Detail of a pervaporation process. A and D evaporate at the gas/liquid interface in the pore and then diffuse across the pore toward the carrier gas. The pore has been drawn straight for illustration purposes only. In a porous membrane, pores are expected to be tortuous.

within the membrane pores and, consequently, diffuse through the membrane independently. Hence, Equation 3.75 describes the fluxes of A and D in the membrane, as long as the appropriate concentrations and binary diffusivities, for example, D_{AB} versus D_{DB}, are used.

Consider the description of the transport of A through the membrane. Let c_{A0} be the concentration of A in the liquid that is in contact with the membrane. Inside the membrane's pores, the gas will be at equilibrium with the liquid. Considering that A is dilute in the liquid, the partial pressure of A in the gas at the gas/liquid interface, P_{A1}, is given by Henry's law, which quantifies gas/liquid equilibrium in dilute systems,

$$P_{A1} = H_A c_{A0} \tag{3.82}$$

where H_A (Henry's law constant) depends on temperature. For an ideal gas, the molar concentration is related to the partial pressure by the ideal gas equation,

$$c_A = \frac{P_A}{RT} \tag{3.83}$$

so that Equation 3.75 can be written for the transport of A through the membrane as follows:

$$N_A = \frac{D_{AB}}{RT}\frac{(P_{A1} - P_{A2})}{\delta} \tag{3.84}$$

where δ is the membrane thickness. Using Equation 3.82 we can state

$$N_A = \frac{D_{AB}}{RT}\frac{(H_A c_{A0} - P_{A2})}{\delta} \tag{3.85}$$

A similar equation applies for species D,

$$N_D = \frac{D_{DB}}{RT}\frac{(H_D c_{D0} - P_{D2})}{\delta} \tag{3.86}$$

If the flow of the carrier gas over the membrane is fast enough, A and D are swept away as they emerge from the membrane, which would allow us to assume that $P_{A2} \approx P_{D2} \approx 0$. The effectiveness of the separation scheme depends on the relative values of the two fluxes, N_A and N_D. Equations 3.85 and 3.86 may be used to quantify this effectiveness and design the process. The ratio of the fluxes is a quantitative indication of the separation effectiveness:

$$\frac{N_A}{N_D} = \frac{D_{AB} H_A c_{A0}}{D_{DB} H_D c_{D0}} \tag{3.87}$$

For example, to enhance the flux of A over D, we would need one or more of the following:

1. A higher value of Henry's constant.
2. A higher diffusivity.

3.3.3 Evaporation of a Liquid into a Stagnant Gas Layer: The Stefan Problem

We investigate an archetypal problem that involves diffusion in the gas phase. Consider the arrangement depicted in Figure 3.9, where a pure liquid A evaporates into an open tube. A purge gas B flows across the mouth of the tube, sweeping away molecules of A as they emerge from the opening. Slowly over time, the liquid level decreases and L, as defined in Figure 3.9, increases.

The problem for L(t) belongs to a broad class of moving-boundary problems known as Stefan problems (Rubinstein, 1971; Gupta, 2003), and the basic experimental set-up is referred to as a Stefan tube. When matched to the solution of the Stefan problem, the Stefan tube becomes a standard apparatus to measure binary gas-phase diffusivities. Owing to this connection between analysis and experiment, our reasons for studying the Stefan problem extend beyond an investigation of one-dimensional, gas-phase diffusion; we also hope to illustrate that, by application of basic principles of mass transport, we can relate that which we can see and measure—for example, the change in the liquid level—to processes that we cannot perceive—for example, migration of species A upwards through the Stefan tube.

The mathematical formulation that we present hereafter incorporates several simplifying approximations. The most important of these is the notion of a pseudo-steady state. Also known as a quasi-steady state, the concept is often useful in problems that

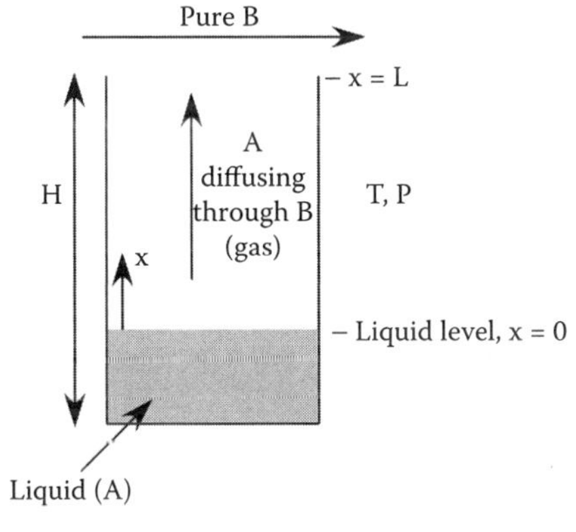

FIGURE 3.9 Evaporation of a liquid (A) into a stagnant gas (B) in a tube. Species A evaporates from the surface of the liquid and moves upwards by diffusion. At the top of the tube, a flow of dry gas carries away the evaporated liquid. The process is carried out at constant temperature and pressure.

involve a significant discrepancy in the time scales of two or more salient rate processes. In the Stefan problem there is a disparity in the rate at which the liquid level falls versus the rate at which species A migrates through the gas phase. As long as liquid A is neither at nor near the boiling point, L will increase slowly compared to the rate of gas-phase mass transport. For the time being, then, we simply assume that the gas-phase fluxes and concentrations of A and B are approximately those that would arise if L were in fact constant. Later, toward the end of the analysis, we will assess the validity of this approximation and show that our quasi-steady analysis works because the molar concentrations of the vapor and liquid phases differ by orders of magnitude.

Additional constraints that we stipulate for the analysis are

1. The temperature and pressure of the system are constant and uniform. Evaporation requires thermal energy and thus the liquid will tend to cool down as it evaporates, but if heat transfer from the surroundings is sufficiently rapid, the liquid temperature will remain approximately constant.
2. The purge gas B flowing across the top of the Stefan tube contains essentially no A. The assumption here is that the flow is fast enough that all the A that reaches the top of the tube is removed but slow enough that its motion does not cause appreciable mixing inside the tube. Thus all motion inside the tube occurs as a consequence of diffusion.
3. Species B is insoluble in liquid A.

To develop an expression for L(t), we will first apply the relations derived for steady, one-dimensional mass transport to work out the gas-phase fluxes of species A and B. We then write a transient mass balance on the liquid phase by relating the change in the liquid level to the gas-phase flux of species A. We solve the mass balance for L(t) and subsequently derive an equation that expresses D_{AB} in terms of measurable changes in L.

We begin our analysis of transport in the Stefan tube by looking at the flux of B. Species A and B do not react in the gas phase, so Equation 3.66 implies

$$N_B = \text{constant}, \; 0 \le x \le L \tag{3.88}$$

Note that Equation 3.66 corresponds to a steady-state process, so that we are invoking the pseudo-steady-state approximation discussed above. Since B is insoluble in the liquid, there is no mass transfer of B through the gas/liquid interface, which implies that $N_B = 0$ at $x = 0$. Equation 3.88 then implies

$$N_B = 0, \; 0 \le x \le L \tag{3.89}$$

Hence, the evaporating species A diffuses through a stagnant column of gas B. Similarly, Equation 3.65 implies that the molar flux of A, N_A, is a constant.

Consider now the one-dimensional flux Equation 3.67. Using $N_B = 0$ and $x_A = c_A/c$, and noticing that c (=P/RT for an ideal gas) is uniform yields

$$N_A\left(1-\frac{c_A}{c}\right) = -D_{AB}\frac{dc_A}{dx} \tag{3.90}$$

This equation can be separated and integrated as follows:

$$N_A\int_0^L dx = -cD_{AB}\int_{c_{A0}}^{c_{AL}}\frac{dc_A}{c-c_A} \tag{3.91}$$

where c_{A0} and c_{AL} are the concentrations of A in the gas at the liquid surface ($x = 0$) and the top of the tube ($x = L$), respectively. Integrating yields

$$N_A = \frac{cD_{AB}}{L}\ln\frac{(c-c_{AL})}{(c-c_{A0})} \tag{3.92}$$

This equation can also be expressed in terms of partial pressures using the ideal gas equation

$$c_A = \frac{P_A}{RT}, \quad c = \frac{P}{RT} \tag{3.93}$$

which yields

$$N_A = \frac{D_{AB}P}{LRT}\ln\frac{(P-P_{AL})}{(P-P_{A0})} \tag{3.94}$$

In due course, we will want to relate the flux of species A to the (slow) change in the liquid level with respect to time, L(t). However, the partial pressures P_{AL} and P_{A0}, which appear in Equation 3.94, are not typically measured as part of the Stefan tube experiment. Instead, two simplifications are introduced. First, the purge gas sweeping across the top of the tube is pure B, so $P_{AL} = 0$. Also, because the liquid evaporates slowly, vapor–liquid equilibrium should prevail locally at the surface of the liquid ($x = 0$).

The equilibrium between a pure liquid and a mixed vapor when the vapor phase can be assumed to be an ideal gas and the liquid is an ideal liquid is described by Raoult's law (e.g., see Felder and Rousseau, 2005). Raoult's law states that the partial pressure in the gas phase of the evaporating species is the saturation (vapor) pressure of the species at the systems temperature. In this case, we can state $P_{A0} = P_{Asat}(T)$, where T is the temperature in the tube and P_{Asat} is the vapor pressure of A at that temperature. These approximations reduce Equation 3.94 to

$$N_A = -\frac{PD_{AB}}{LRT}\ln\left(1-\frac{P_{Asat}}{P}\right) \tag{3.95}$$

The partial pressure of species A changes with position in the tube. The profile, $P_A(x)$, can be found by integration of Equation 3.102 from the interface to an arbitrary position x. The resultant expression reads

$$N_A \int_0^x dx = -\frac{PD_{AB}}{RT} \int_{P_{Asat}}^{P_A} \frac{dP_A}{P - P_A} \tag{3.96}$$

Evaluation of the integrals yields

$$N_A x = \frac{PD_{AB}}{RT} \ln\left(\frac{P - P_A}{P - P_{Asat}}\right) \tag{3.97}$$

Substitution of Equation 3.95 into Equation 3.97 leads to, after manipulations,

$$\frac{P_A}{P} = 1 - \left(1 - \frac{P_{Asat}}{P}\right)^{1-x/L} \tag{3.98}$$

The flux given by Equation 3.95 and the partial pressure profile given by Equation 3.98 correspond to a specific value of L. Under the pseudo-steady-state approximation, Equation 3.95 gives the **instantaneous** evaporation flux that would be measured when the liquid level is equal to the specified value. However, as the liquid evaporates, its level will drop, which implies that L will grow with time until eventually all the liquid evaporates (L = H in Figure 3.9). As L increases, the evaporation flux predicted by Equation 3.95 decreases.

To determine how the liquid level changes with time, we perform a macroscopic mass balance on the liquid contained in the Stefan tube. Let ρ_l and V_l, respectively, denote the density and volume of liquid, so that the total mass of liquid at a given time is given by $\rho_l V_l$. A mass balance following the approach exposed in Chapter 2 reads

$$\frac{d(\rho_l V_l)}{dt} = -M_A N_A A \tag{3.99}$$

where A is the cross-sectional area of the Stefan tube. The volume of liquid is related to the geometry of the tube by

$$V_l = A\,(H - L) \tag{3.100}$$

so that

$$\frac{dV_l}{dt} = -A\frac{dL}{dt} \tag{3.101}$$

Substituting into Equation 3.99 and rearranging yields

$$\frac{dL}{dt} = \frac{M_A}{\rho_l} N_A \tag{3.102}$$

Using Equation 3.95 for the flux leads to

$$\frac{dL}{dt} = -\frac{M_A D_{AB} P}{\rho_l LRT} \ln\left(1 - \frac{P_{Asat}}{P}\right) \tag{3.103}$$

On the right-hand side of this equation, the only quantity that changes with time is L. Therefore, we can separate and integrate the equation as follows:

$$\int_{L_0}^{L} LdL = -\frac{M_A D_{AB} P}{\rho_l RT} \ln\left(1 - \frac{P_{Asat}}{P}\right) \int_0^t dt \tag{3.104}$$

where L_0 is the initial ($t = 0$) distance from the liquid surface to the top of the tube. Integrating yields an expression to calculate L as a function of time:

$$L(t) = L_0 \sqrt{1 - 2t \frac{M_A D_{AB} P}{\rho_l RT L_0^2} \ln\left(1 - \frac{P_{Asat}}{P}\right)} \tag{3.105}$$

As mentioned before, the Stefan tube experiment can be used in practice to measure gas-phase diffusivities. In the experiment, a liquid of known density, molecular weight, and vapor pressure is placed in the tube and L_0 is measured. After a time t_1, the new length of tube occupied by the gas is measured, L_1. From Equation 3.105, we obtain the following equation to calculate the diffusivity from the data,

$$D_{AB} = -\frac{1}{2} \frac{\rho_l RT}{M_A P} \frac{L_1^2 - L_0^2}{t_1} \left[\ln\left(1 - \frac{P_{Asat}}{P}\right) \right]^{-1} \tag{3.106}$$

Alternatively, a more accurate approach would be to generate a data set of L vs. t and then fit Equation 3.105 to the data using the diffusivity as an adjustable parameter.

Another application of the analysis is the calculation of the time that it takes for all the liquid in the tube to evaporate, t_f. This can be done by letting $L = H$ in Equation 3.105 and solving for t_f,

$$t_f = -\frac{(H^2 - L_0^2)\rho_l RT}{2 M_A D_{AB} P} \left[\ln\left(1 - \frac{P_{Asat}}{P}\right) \right]^{-1} \tag{3.107}$$

Having solved the problem, we can now assess the validity of the pseudo-steady-state approximation. For this purpose, we compare v_A, the species velocity at which A diffuses through the gas phase, to dL/dt, the speed at which the gas/liquid interface falls. Combining Equations 3.5 and 3.102 clearly shows that dL/dt is proportional to v_A, namely,

$$\frac{dL}{dt} = \frac{M_A c_A}{\rho_l} v_A \tag{3.108}$$

The coefficient that determines the relative magnitudes of the two velocities is $M_A c_A / \rho_l$, which is the ratio between the molar concentration of A in the gas phase and the molar concentration of the liquid. Gases are ordinarily far less dense than liquids, so we expect $M_A c_A / \rho_l \ll 1$. Under this condition, Equation 3.108 clearly shows that the rate at which the interface drops in the tube would be much slower than the rate at which A diffuses in the gas phase which, in turn, means that steady state in the diffusion process will be achieved before the interface drops appreciably.

To gain a quantitative feel for the quantity $M_A c_A / \rho_l$, let us consider typical values for benzene, a volatile organic liquid. The molar mass of benzene is 78.1 g/mol. The density and vapor pressure of the liquid are 0.879 g/cm^3 and 14 kPa, respectively, at 20°C. If A is benzene, then $\rho_l / M_A = 1.13 \times 10^{-2}$ mol/cm^3 is the molar density of the liquid. In the gas phase, c_A will be highest in the neighborhood of the gas/liquid interface, meaning $c_A \le P_{Asat} / RT$ throughout the gas phase in the tube ($0 \le x \le L$). The vapor pressure of benzene translates into $P_{Asat} / RT = 5.74 \times 10^{-6}$ mol/cm^3 and we obtain

$$\frac{M_A c_A}{\rho_l} \le 5.08 \times 10^{-4} \ll 1 \tag{3.109}$$

Clearly, the pseudo-steady approximation would be applicable to the evaporation of benzene in a Stefan tube at 20°C.

Another way to think about the quasi-steady approximation is in terms of time scales. The pseudo-steady state arises because the time scale on which the liquid level changes is long compared to the time scale on which species A migrates out of the Stefan tube. Estimating these time scales is a straightforward exercise. The height of the vapor phase is L, so species A migrates out of the Stefan tube in a time that is on the order of L/v_A. The liquid level, on the other hand, changes in a time frame given by L/(dL/dt); this estimate is based on the idea that motion of the interface is only significant when changes in L are comparable to L. The quasi-steady approximation is valid if L/vA ≪ L/(dL/dt), which is equivalent to saying dL/dt ≪ v_A.

To illustrate an application of the preceding analysis, consider the following situation. Gasoline has spilled onto the ground by accident and filtered down through the unsaturated soil to a depth of 3 m, where it has formed a liquid pool with a thickness of 2 cm. We wish to estimate the time required for the entire gasoline pool to evaporate. As a worst-case scenario, consider that the evaporated gasoline travels through the air in the soil pores by diffusion only. If we model the soil as an assembly of parallel pores, then the process corresponds to the Stefan problem. For the purpose of this application, let T = 25°C, P = 1 atm, and the gasoline is modeled as consisting of pure n-octane. Relevant physical properties of n-octane at these conditions (Felder and Rousseau, 2005; Geankoplis, 1993) are: $M_A = 114.22$ g/mol, $\rho_l = 0.703$ g/cm^3, $P_{Asat} = 0.0184$ atm, and $D_{AB} = 0.08$ cm^2/s. Equation 3.107 can be evaluated to obtain $t_f = 701$ days for the specified lengths (H = 3 m, $L_0 = 2.98$ m). More importantly, the result of this calculation gives the reader an idea of how slow the recovery of a contamination event, such as the one considered here, is when the remediation process is controlled by diffusion.

The solutions to problems presented heretofore show how the principle of mass conservation, coupled with the one-dimensional version of Fick's law, can be used to solve relatively simple mass transport problems. However, most practical problems have a higher degree of complexity, involving species concentration fields that may be three dimensional in form and time dependent by nature. Therefore, it is necessary to have a more comprehensive description of mass transport in continua. In the next chapter, we will develop equations that govern concentration fields and fluxes in terms that are more suited for general application. The problems investigated in the foregoing section can be solved with these general equations, as an alternative to the approach that we have used to this stage.

PROBLEMS

3.1 A container is filled with pure oxygen gas at 1 atm and 25°C. Oxygen leaks from the container to the environment (atmospheric air at 1 atm and 25°C, 21 mol% O_2) due to diffusion through the container wall, which is made of polyethylene (PE) with a thickness $\delta_1 = 0.3$ cm. To reduce the leak, it is proposed that a thin layer of latex coating (LC) be applied to the exterior of the container (see figure). If the maximum tolerable leak flux is 1.5×10^{-13} mol O_2/cm^{2}s at steady state, determine the minimum thickness of the latex coating. Assume that mass transfer between the surface of the latex coating and atmospheric air is very fast. Additional information:

$$D_{O_2,PE} = 1.7 \times 10^{-7}\,\frac{cm^2}{s}$$

$$D_{O_2,LC} = 1.0 \times 10^{-8}\,\frac{cm^2}{s}$$

The solubility of O_2 in each solid is given by

$$c_{O_2} = K_i P_{O_2}, \quad \text{with:} \quad K_{PE} = 1.2 \times 10^{-6}\,\frac{mol}{cm^3 atm}, \quad K_{LC} = 6.5 \times 10^{-7}\,\frac{mol}{cm^3 atm}$$

Assume that the equilibrium relation at the PE/LC interface is given by

$$\frac{c_{O_2,PE}}{c_{O_2,LC}} = \frac{K_{PE}}{K_{LC}}$$

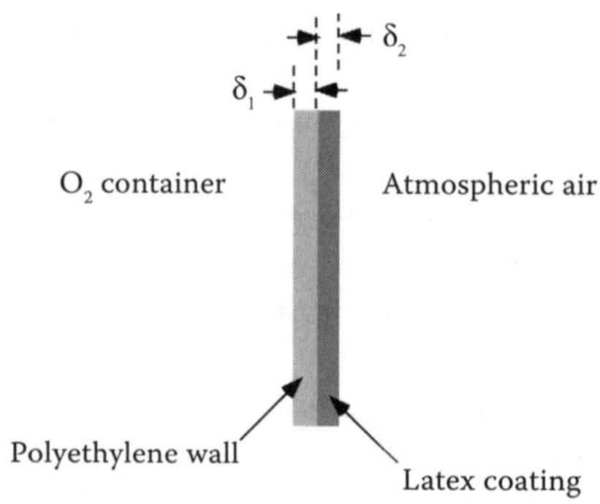

3.2 Steady-state experiments have been used to measure the ratio between the molar flux of oxygen (N_A) and the oxygen partial pressure difference (ΔP_A) due to diffusion across a 2 mm-thick polymeric flat film exposed to gas (T = 25°C) containing oxygen on both sides. The result is

$$\frac{N_A}{\Delta P_A} = 4 \times 10^{-8} \frac{\text{mol}}{\text{m}^2\text{s atm}}$$

At equilibrium, the concentration of oxygen in the film polymer is directly proportional to its partial pressure in the gas.

A material that is susceptible to oxidation is packaged in a container that has a gas volume of 150 cm³. To avoid oxidation, the inside of the container is flushed with pure nitrogen so that the gas inside contains initially no oxygen. The package has a window made of the polymer described above with area equal to 5 cm² and thickness of 1 mm, which separates the gas inside from atmospheric air (T = 25°C, P = 1 atm, P_A = 0.21 atm). Assuming that the diffusion process through the polymer window is at steady state at all times, find the time required for the oxygen partial pressure to rise to 0.01 atm inside the package. Assume that total gas pressure inside the package is constant at 1 atm and that the gas within the package is well mixed at all times.

3.3 Gaseous mixtures of ammonia (A) and nitrogen (B) are contained in two tanks connected by a glass tube 61 cm long with an inside diameter of 2.44 cm (see figure) at T = 298 K and P = 101.32 kPa (both constant). The partial pressures of ammonia in the two tanks are 20.0 and 6.666 kPa. The tanks are so large that these partial pressures can be considered constant, even though ammonia will diffuse through the tube from tank 1 to tank 2. Since an imbalance of the pressure between the two tanks would lead to a hydrodynamic flow through the tube, the total number of moles in each tank will remain constant. This means that nitrogen will diffuse from tank 2 to tank 1 at the same molar rate as ammonia is diffusing from tank 1 to tank 2. This process is called **equimolar counterdiffusion**. Assume that mass transfer through the tube is at steady state. The diffusivity for this system is 2.3×10^{-5} m²/s.

a. Calculate the molar flow rates of ammonia and nitrogen through the tube in kmol/s.
b. Calculate the partial pressures of ammonia and nitrogen at the midpoint in the tube (0.305 m from each chamber). Plot the partial pressure profile of ammonia in the tube.
c. Calculate the species velocities, v_A and v_B at both ends of the tube.
d. Calculate the mass average velocity of the mixture, v, at both ends of the tube. Comment on your results.

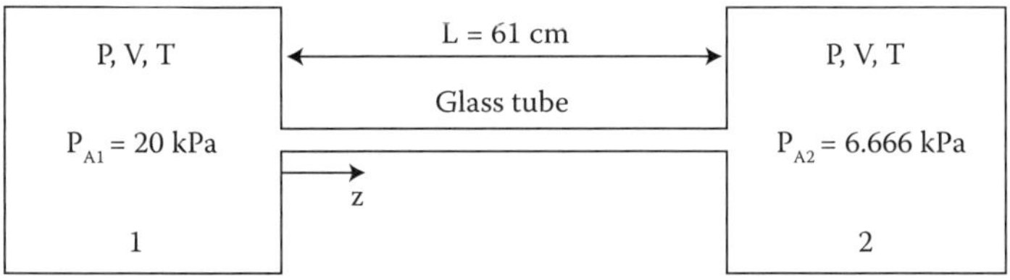

3.4 Someone has left 10 cm^3 of liquid trichloroethylene (TCE) in a graduated cylinder in an unventilated, closed laboratory on Friday at 4 pm. The graduated cylinder has an internal diameter of 3 cm and a length of 35 cm. The lab has a total volume of 600 m^3 and the air inside it will be stagnant during the weekend. The EPA has established an indoor standard for TCE of 100 μg/m^3 as the maximum allowable concentration of TCE in air for a typical 8-h exposure. Estimate the concentration of TCE in the lab air on Monday at 8 am and establish if it is safe to go into the lab unprotected. The density of liquid TCE is 0.703 g/cm^3; its vapor pressure at 25°C is 0.0184 atm and its molecular weight is 114.2 g/mol. The diffusivity of TCE in air is 0.08 cm^2/s. Consider that the atmospheric pressure is 0.91 atm.
Hint: Assume that the air in the lab is well mixed at all times, so that the process is controlled by the diffusion of TCE in the graduated cylinder. Why would this be a good assumption to use to assess safety?

4 Diffusive Transport

Applications of the Diffusion Equation

4.1 CONSERVATION OF MASS OF A CHEMICAL SPECIES IN A MIXTURE

In Section 2.2, we derived the mathematical form of the macroscopic mass balance for a species in a mixture (Equation 2.39). In our analysis, we were not concerned with processes occurring at each point of the control volume but rather with an average view of mass transfer. That approach allowed us to formulate and solve problems in perfectly mixed systems, such as the CSTR. However, in many practical applications, it is necessary to quantify mass transfer at each point in the continuum. In this section we will derive the general **point equations** that represent species mass conservation. We start by writing an instantaneous species mole balance in a control volume along the lines of Equation 2.32. However, we will apply this mole balance to a control volume (Figure 4.1) that is a small parallelepiped of size $\Delta x \times \Delta y \times \Delta z$ with the intention of eventually letting Δx, Δy, $\Delta z \to 0$, to find an equation that is applicable at a point in the continuum. The species mole balance is

$$\left\{\begin{matrix}\text{The time rate of change of the}\\ \text{total moles of A in a control volume}\end{matrix}\right\} = \left\{\begin{matrix}\text{The } \textbf{net} \text{ rate at which moles of}\\ \textbf{A enter} \text{ the control volume}\end{matrix}\right\} + \left\{\begin{matrix}\text{The } \textbf{net} \text{ rate at which moles}\\ \text{of A are } \textbf{created} \text{ in the control}\\ \text{volume by chemical reactions}\end{matrix}\right\} \tag{4.1}$$

Since the control volume is arbitrarily small, we will consider that the species molar concentration is uniform and, therefore, represented by the single value c_A. In view of this, the accumulation terms in the balance can be written as follows:

$$\left\{\begin{matrix}\text{The time rate of change of the}\\ \text{total moles of A in a control volume}\end{matrix}\right\} = \frac{\partial}{\partial t}(c_A \Delta x \Delta y \Delta z) \tag{4.2}$$

Input and output of species A can occur only across the boundaries of the control volume, represented by the six faces. The moles per unit time that cross each of the

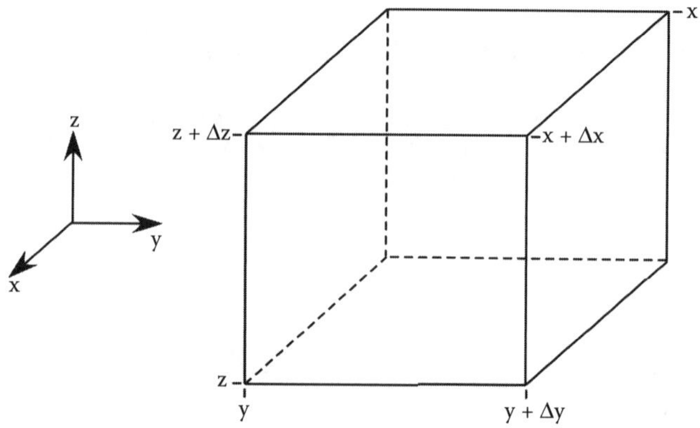

FIGURE 4.1 Control volume of differential size used in the derivation of point equations.

parallelepiped's faces are equal to the corresponding component of the molar flux vector, $\mathbf{N}_A$, times the surface area. According to Figure 4.2, we can write

$$\begin{Bmatrix} \text{The } \textbf{net} \text{ rate at which moles of} \\ \textbf{A enter} \text{ the control volume} \end{Bmatrix} = N_{Ax}\Big|_x \Delta y \Delta z - N_{Ax}\Big|_{x+\Delta x} \Delta y \Delta z + N_{Ay}\Big|_y \Delta x \Delta z$$
$$-N_{Ay}\Big|_{y+\Delta y} \Delta x \Delta z + N_{Az}\Big|_z \Delta x \Delta y - N_{Az}\Big|_{z+\Delta z} \Delta x \Delta y \tag{4.3}$$

Note that components of the flux entering the control volume are positive whereas components that represent moles of A leaving the volume are negative.

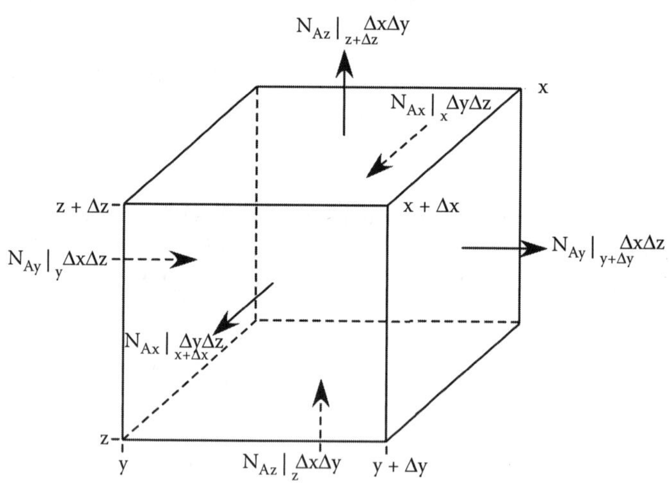

FIGURE 4.2 Molar flow rates of species A that cross the boundaries of the control volume.

To quantify the reaction term in Equation 4.1, we use the reaction rate R_A, which represents the net rate of A created per unit time and volume,

$$\left\{\begin{array}{l}\text{The } \textbf{net} \text{ rate at which moles} \\ \text{of A are } \textbf{created} \text{ in the control} \\ \text{volume by chemical reactions}\end{array}\right\} = R_A \Delta x \Delta y \Delta z \tag{4.4}$$

Substituting Equations 4.2 to 4.4 into Equation 4.1 leads to

$$\begin{aligned}\frac{\partial}{\partial t}(c_A \Delta x \Delta y \Delta z) = {} & N_{Ax}\big|_x \Delta y \Delta z - N_{Ax}\big|_{x+\Delta x} \Delta y \Delta z + N_{Ay}\big|_y \Delta x \Delta z - N_{Ay}\big|_{y+\Delta y} \Delta x \Delta z \\ & + N_{Az}\big|_z \Delta x \Delta y - N_{Az}\big|_{z+\Delta z} \Delta x \Delta y + R_A \Delta x \Delta y \Delta z\end{aligned} \tag{4.5}$$

The next step is to divide this equation by Δx Δy Δz and take the limit as Δx, Δy, Δz → 0. This leads to

$$\frac{\partial c_A}{\partial t} = -\frac{\partial N_{Ax}}{\partial x} - \frac{\partial N_{Ay}}{\partial y} - \frac{\partial N_{Az}}{\partial z} + R_A \tag{4.6}$$

The first three terms on the right-hand side of this equation are the divergence of the molar flux vector, so that

$$\frac{\partial c_A}{\partial t} + \nabla \cdot \mathbf{N}_A = R_A \tag{4.7}$$

This is the **continuity equation of species A**. At this point, the equation contains both the concentration and the flux vector as dependent variables. To complete the formulation, we will consider first the special case of a binary system with components A and B (the final equations also will be directly applicable to pseudo-binary systems, as defined in Section 3.2). For this case, the molar flux of A is given by (Equations 3.37 and 3.44)

$$\mathbf{N}_A = x_A \mathbf{N} - c D_{AB} \nabla x_A \tag{4.8}$$

Multiplying Equation 3.9 by x_A, we can write

$$x_A \mathbf{N} = c_A \mathbf{v}^* \tag{4.9}$$

We will restrict our analysis to systems in which the molar density, c, can be considered uniform. Using this assumption and substituting Equation 4.9 into Equation 4.8 leads to

$$\mathbf{N}_A = c_A\mathbf{v}^* - D_{AB}\nabla c_A \tag{4.10}$$

Substitution of this equation into the continuity Equation 4.7 yields

$$\frac{\partial c_A}{\partial t} + \nabla\cdot(c_A\mathbf{v}^*) = \nabla\cdot(D_{AB}\nabla c_A) + R_A \tag{4.11}$$

If we assume that D_{AB} is uniform, we get

$$\frac{\partial c_A}{\partial t} + \nabla\cdot(c_A\mathbf{v}^*) = D_{AB}\nabla^2 c_A + R_A \tag{4.12}$$

The second term on the left-hand side of this equation represents the convective transport of A. The molar average velocity, $\mathbf{v}^*$, is, in general, different from the mass average velocity, $\mathbf{v}$, as argued in Section 3.2. However, most applications of mass transfer will be for cases in which the species whose transport is quantified (A) is present in dilute conditions (this would be consistent with the assumptions made before of constant c and D_{AB}). If this is the case, then $x_A \ll 1$ and, consequently, $x_B \approx 1$, so that

$$\mathbf{v}^* \approx \mathbf{v}_B \approx \mathbf{v} \tag{4.13}$$

and $\mathbf{v}^*$ can be replaced by $\mathbf{v}$ in Equation 4.12. In addition, for incompressible flows (see Appendix B), the divergence of the velocity vector is zero,

$$\nabla\cdot \mathrm{v} = 0 \Rightarrow \nabla\cdot(c_A\mathbf{v}) = \mathbf{v}\cdot\nabla c_A + c_A\nabla\cdot\mathbf{v} = \mathbf{v}\cdot\nabla c_A \tag{4.14}$$

and Equation 4.12 becomes

$$\frac{\partial c_A}{\partial t} + \mathbf{v}\cdot\nabla c_A = D_{AB}\nabla^2 c_A + R_A \tag{4.15}$$

This is the **convective–diffusion equation**, which is widely used to represent mass transfer problems. As developed, the equation is valid for binary systems; that is, mixtures of two species A and B. However, the equation is also applicable to the transport of A in a multicomponent mixture in which A is dilute. In that case, the binary diffusivity, D_{AB}, is replaced by the diffusivity of A in the mixture, D_A,

$$\frac{\partial c_A}{\partial t} + \mathbf{v}\cdot\nabla c_A = D_A\nabla^2 c_A + R_A \tag{4.16}$$

This equation, along with expressions of the species molar flux, is presented in expanded form for various coordinate systems in Appendix C.

For the particular case in which the material is not moving: $\mathbf{v} \equiv 0$, Equation 4.16 is simplified by eliminating the convective term. This is the **diffusion equation**. In this chapter, we consider applications of the diffusion equation.

4.2 ONE-DIMENSIONAL, STEADY DIFFUSION

The problems to be solved in this section correspond to binary systems or multicomponent systems in which the species whose transport is being studied (A) is present at dilute conditions ($x_A \ll 1$), which makes Fick's law applicable. For binary systems, the approach to a specific problem will depend on whether the mixture is dilute in A or not:

1. For dilute systems, Equation 4.16 with $\mathbf{v} = 0$ applies. Note that taking $\mathbf{v} = \mathbf{0}$ is really an approximation, since A will be diffusing and hence there will be motion in the material. However, bulk motion will be negligible due to the low concentrations of A. In terms of the representation of the molar flux of A, we can see that, since $x_A \ll 1$, we can state that

$$\mathbf{N}_A \approx \mathbf{J}_A^* \tag{4.17}$$

   and the components of the molar flux of A can be directly found from Fick's law.
2. For concentrated binary systems, the continuity equation in terms of the molar flux (Equation 4.7) must be used, along with the general form of the flux equation (Equation 4.8). For these problems, information will be needed on both components (i.e., if solving for concentration profiles and flux of A, we will need to have information on the flux of B). This is the approach that was used in Section 3.3.3 to solve the Stefan problem.

4.2.1 STEADY DIFFUSION THROUGH AN ANNULAR MEMBRANE

In this section, we consider the diffusion of a species (A) through a cylindrical solid membrane made of a species B that separates two fluids with different concentrations of A (Figure 4.3). The conditions and assumptions for this application are

1. Steady state.
2. No chemical reactions in the membrane, $R_A = 0$.
3. Species A is dilute in the membrane ($x_A \ll 1$).
4. Symmetry in both z and θ directions allows us to state that the concentration will depend only on radial direction: $c_A = c_A(r)$ in the membrane.
5. Within the membrane, c and D_{AB} are uniform.
6. The fluid inside the cylinder is the same as outside the cylinder, and the concentration of A in the fluid is uniform (c_{A1} and c_{A2} are uniform in the fluid phases).
7. At the fluid/solid interface, local equilibrium implies that the concentration of A in the membrane is at equilibrium with the concentration of A in the

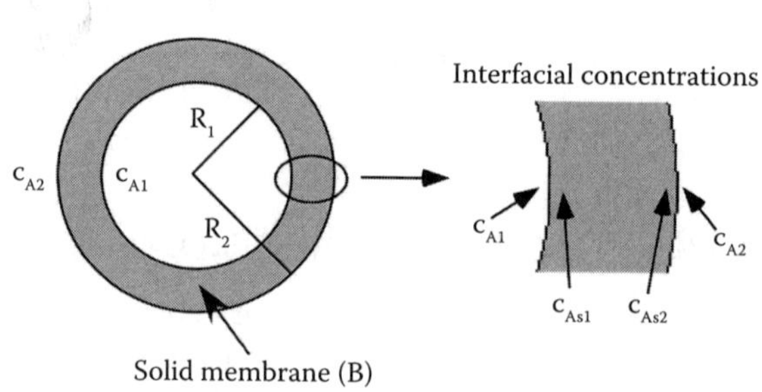

FIGURE 4.3 A long annular membrane separates two fluids with different concentrations of A (c_{A1} and c_{A2}). Equilibrium is assumed at the fluid/solid interfaces so that c_{Asi} (i = 1, 2) represents the concentration in the solid that is at equilibrium with the corresponding fluid concentration, c_{Ai}.

liquid. We will assume that the equilibrium relation is linear, so that it is represented by the equation

$$c_{As} = Kc_{Af} \tag{4.18}$$

where c_{As} is the concentration of A in the membrane that would be at equilibrium with a concentration of A in the fluid c_{Af}, and the partition coefficient, K, is a constant.

Under the conditions listed above, we can state that $c_A = c_A(r)$; that is, the concentration of A depends on radial position only. Furthermore, since $\mathbf{v} = \mathbf{0}$ (no convection) and $R_A = 0$ (no chemical reaction), Equation C.12, Appendix C, simplifies to

$$\frac{d}{dr}\left(r\frac{dc_A}{dr}\right) = 0 \tag{4.19}$$

Integrating this equation twice leads to the solution

$$c_A = A\ln r + B \tag{4.20}$$

where A and B are integration constants. To determine these two constants, we will use the boundary conditions (see Figure 4.3):

$$c_A = c_{As1}, r = R_1 \tag{4.21}$$

$$c_A = c_{As2}, r = R_2 \tag{4.22}$$

Using these two conditions to find the integration constants yields the concentration profile:

$$c_A = c_{As1} - (c_{As1} - c_{As2})\frac{\ln(r/R_1)}{\ln(R_2/R_1)} \tag{4.23}$$

The concentrations in the membrane can be related to the fluid concentrations by the equilibrium relation (4.18), which implies

$$c_{As1} = Kc_{A1} \tag{4.24}$$

$$c_{As2} = Kc_{A2} \tag{4.25}$$

Using these relations, Equation 4.23 can be written as follows:

$$c_A = Kc_{A1} - K(c_{A1} - c_{A2})\frac{\ln(r/R_1)}{\ln(R_2/R_1)} \tag{4.26}$$

Since this is a dilute system, the molar flux of A in the membrane is given by Fick's law,

$$N_{Ar} = -D_{AB}\frac{dc_A}{dr} \tag{4.27}$$

Substitution of Equation 4.26 yields, after manipulations,

$$N_{Ar} = \frac{D_{AB}K}{r}\frac{(c_{A1} - c_{A2})}{\ln(R_2/R_1)} \tag{4.28}$$

This equation shows that the molar flux of A is inversely proportional to r, which is due to the increase in surface area of a cylindrical shell with r. At any radial position, we can calculate the molar flow rate of A (moles of A transported per unit time) in the radial direction by

$$W_{Ar} = 2\pi rLN_{Ar} \tag{4.29}$$

where L is the length of the cylinder. This yields

$$W_{Ar} = 2\pi LD_{AB}K\frac{(c_{A1} - c_{A2})}{\ln(R_2/R_1)} \tag{4.30}$$

which is independent of r. This is an expected result: mass conservation at steady state (no accumulation) implies that the moles of A per unit time being transported across the membrane must be the same at each point.

The solution found assumes that the fluids on both sides of the membrane are perfectly mixed; that is, mass transfer within the fluids is fast enough that concentrations within them are uniform (c_{A1} and c_{A2}). If this is not the case and mass transfer by convection in the fluids is slow enough that concentration gradients can be established, the concentration of A at the fluid/membrane interface will be different from the bulk concentration in the fluid. The concentration profile for the latter case is illustrated in Figure 4.4 for $c_{A1} > c_{A2}$. Note that the fact that the concentrations of A at the fluid/membrane interfaces (c_{Ai1} and c_{Ai2}) are at equilibrium with those in the membrane leads to a discontinuity of the concentration at the interface.

In this case, Equation 4.30 still applies, but the molar flux of A is expressed in terms of the interfacial fluid concentrations depicted in Figure 4.4,

$$W_{Ar} = 2\pi L D_{AB} K \frac{(c_{Ai1} - c_{Ai2})}{\ln(R_2/R_1)} \tag{4.31}$$

The interfacial concentrations, c_{Ai1} and c_{Ai2}, are generally unknown at this point, so that Equation 4.31 cannot be used to evaluate the rate of mass transfer of A through the membrane. The values of these concentrations depend on the mass transfer processes that occur in the fluid phases. Hence, the transport of A through the membrane cannot be quantified separately from its transport in the adjacent fluids. However, the convective transport species flux in a fluid adjacent to a phase interface can be represented in most cases as directly proportional to the concentration difference between the interface and the bulk of the fluid far from the interface. The proportionality constant is the **convective mass transfer coefficient**. Here, we represent the fluxes at the interfaces using convective mass transfer coefficients, k_{c1} and k_{c2},

$$N_{Ar}\big|_{r=R_1} = k_{c1}(c_{A1} - c_{Ai1}) \tag{4.32}$$

$$N_{Ar}\big|_{r=R_2} = k_{c2}(c_{Ai2} - c_{A2}) \tag{4.33}$$

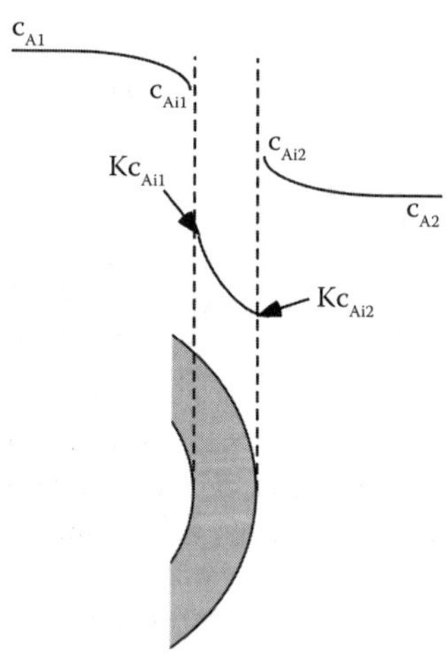

FIGURE 4.4 Shape of the concentration profiles when there is transport by convection in the fluids inside and outside the membrane.

A more general treatment of convection will be used in Chapter 6 to show why this type of expression is appropriate to represent convective mass transfer fluxes. For now, it suffices to say that the mass transfer coefficients depend on the state of motion of the fluid. Hence, the values of k_{c1} and k_{c2} represent the convective transport in the fluids and can be found by either solving the convection problem in the moving fluids, or using empirical correlations based on the particular system under study. Note that, since we are representing the radial component of the flux vector, the concentration difference is defined in the direction of increasing radial coordinate.

From Equations 4.32 and 4.22, we can find the molar flow rate of A, which is the same everywhere:

$$W_{Ar} = 2\pi R_1 L k_{c1}(c_{A1} - c_{Ai1}) \tag{4.34}$$

$$W_{Ar} = 2\pi R_2 L k_{c2}(c_{Ai2} - c_{A2}) \tag{4.35}$$

Equations 4.31, 4.34, and 4.35 conform a system of three equations with three unknowns: W_{Ar}, c_{Ai1}, and c_{Ai2}. A practical way to solve these equations is to rewrite them leaving only concentration differences on the right-hand side, as follows:

$$\frac{\ln(R_2/R_1)W_{Ar}}{2\pi L D_{AB} K} = c_{Ai1} - c_{Ai2} \tag{4.36}$$

$$\frac{W_{Ar}}{2\pi R_1 L k_{c1}} = c_{A1} - c_{Ai1} \tag{4.37}$$

$$\frac{W_{Ar}}{2\pi R_2 L k_{c2}} = c_{Ai2} - c_{A2} \tag{4.38}$$

Direct addition of these three equations allows us to eliminate the interfacial concentrations and find an expression to calculate the rate of transport of A,

$$W_{Ar} = \frac{2\pi L(c_{A1} - c_{A2})}{(1/R_1 k_{c1}) + (\ln(R_2/R_1)/D_{AB}K) + (1/R_2 k_{c2})} \tag{4.39}$$

Each term in the denominator of this equation can be interpreted as a mass transfer resistance, since an increase in its value would lead to a decrease in the rate of transport of A. In fact, the original solution of our problem, Equation 4.30, which considered that the fluids were perfectly mixed, is a special case of this result. If the fluids are well mixed, transport between the interface and the bulk is very fast, and the mass transfer coefficient has a very high value. We can see from Equation 4.39 that if

$$k_{c2} >> \frac{D_{AB}K}{R_2 \ln(R_2/R_1)} \tag{4.40}$$

and

$$k_{c1} >> \frac{D_{AB}K}{R_1 \ln(R_2/R_1)} \tag{4.41}$$

then Equation 4.39 reduces to Equation 4.30 as an approximation.

4.2.2 Diffusion from a Sphere Submerged in a Stagnant Fluid: Dissolution of a Spherical Solid Particle

In this section, we consider a solid spherical particle suspended in a liquid in which the solid is slightly soluble. Far from the particle, the fluid is free of solute. The solid (A) dissolves and diffuses away from the particle, since the liquid (B) contains no A far from the particle. The expected concentration profile of A is shown in Figure 4.5. One objective of the analysis will be to determine the rate of dissolution of the solid.

Spherical coordinates will be used to formulate the diffusion problem in the region $r > R$ (see Appendix C for notation). The conditions and assumptions under which this problem will be solved are

1. Symmetry in θ (polar angle) and φ (azimuthal angle).
2. No chemical reactions, $R_A = 0$.
3. Uniform c (liquid molar density) and D_{AB}.
4. Species A is dilute in the liquid, $c_{As} \ll c$, where c is the liquid molar density and c_{As} is the solubility of A in the liquid.
5. This is evidently not a steady-state problem since, as A dissolves, the particle radius will decrease with time. However, if the solid is sparingly soluble in the liquid, the diffusion process in the liquid will approach steady state

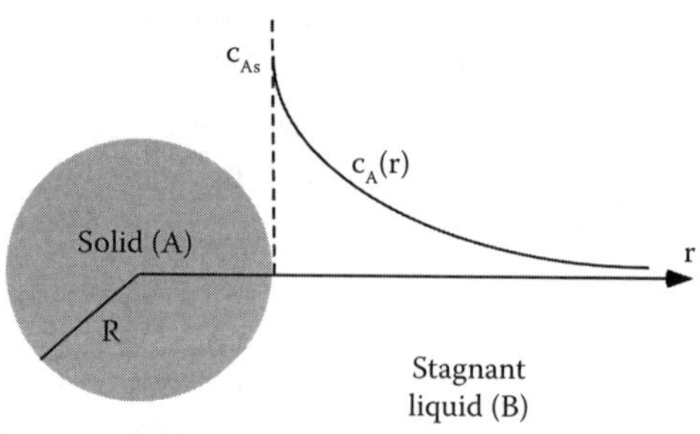

FIGURE 4.5 A solid spherical particle dissolves in a large, stagnant volume of liquid. Under pseudo-steady-state conditions (see text), the concentration of the dissolving species at the surface of the particle will be the solubility of A in B (c_{As}), and it will decay uniformly with radial position as shown.

on a time that is much shorter than that required for appreciable changes in the particle size. In this case, the diffusion process may be assumed to be at steady state even though the process is inherently unsteady. This is called the pseudo-steady-state approximation, which was discussed in detail and applied to the Stefan problem in Chapter 3.

Under the assumptions and conditions listed, the continuity equation in spherical coordinates (Appendix C) simplifies to

$$\frac{d}{dr}\left(r^2 \frac{dc_A}{dr}\right) = 0 \tag{4.42}$$

Integrating this equation twice leads to

$$c_A = \frac{A}{r} + B \tag{4.43}$$

where A and B are integration constants. Appropriate boundary conditions are

$$\begin{aligned} c_A &= c_{As},\ r = R \\ c_A &= 0,\ r \to \infty \end{aligned} \tag{4.44}$$

which yields the concentration profile

$$c_A = c_{As} \frac{R}{r} \tag{4.45}$$

The flux of A from the surface of the particle can be obtained from Fick's law,

$$N_{Ar}\Big|_{r=R} = -D_{AB} \frac{dc_A}{dr}\Big|_{r=R} \tag{4.46}$$

which yields

$$N_{Ar}\Big|_{r=R} = \frac{c_{As} D_{AB}}{R} \tag{4.47}$$

Note that this dissolution flux changes with time: as the sphere shrinks, R decreases, and the flux increases. To determine how the radius of the sphere changes with time, we write a macroscopic mole balance using the solid as control volume, following the approach detailed in Chapter 2,

$$\frac{d(c_s V)}{dt} = -N_{Ar}\Big|_{r=R} 4\pi R^2 \tag{4.48}$$

where the solid molar density, c_s, is a constant. Substitution of Equation 4.47 into Equation 4.48 and using $V = 4\pi R^3/3$ yields

$$\frac{dR}{dt} = -\frac{c_{As} D_{AB}}{R c_s} \tag{4.49}$$

Separating and integrating from $R = R_0$ at $t = 0$ yields

$$\int_{R_0}^{R} R dR = -\frac{c_{As} D_{AB}}{c_s} \int_0^t dt \tag{4.50}$$

which leads to

$$R = \sqrt{R_0^2 - \frac{2 c_{As} D_{AB}}{c_s} t} \tag{4.51}$$

From this equation, we can calculate the time necessary to dissolve the particle completely (t_f). This will be achieved when $R = 0$, which leads to

$$t_f = \frac{R_0^2 c_s}{2 c_{As} D_{AB}} \tag{4.52}$$

The solution found can be used to analyze the effect of particle size on dissolution rate. A critical result is that the time necessary to dissolve the particle completely is proportional to R_0^2. An interesting question to ask at this point is: how would the dissolution time change if we want to dissolve the **same amount** of material but we distribute it in particles of different sizes? Consider that we want to dissolve a given amount of solid, represented by its total volume, V_T. If we form a single sphere with this amount of solid, we would have

$$R_0 = \left(\frac{3V_T}{4\pi}\right)^{1/3} \tag{4.53}$$

and the time to dissolve the particles is (Equation 4.52)

$$t_{f1} = \frac{c_s}{2 c_{As} D_{AB}} \left(\frac{3V_T}{4\pi}\right)^{2/3} \tag{4.54}$$

Now suppose that we distribute the same volume into two equal particles. Their radii are

$$R_0 = \left(\frac{3V_T}{8\pi}\right)^{1/3} \tag{4.55}$$

and the time to dissolve each of them is

$$t_{f2} = \frac{c_s}{2c_{As}D_{AB}}\left(\frac{3V_T}{8\pi}\right)^{2/3} \tag{4.56}$$

Dividing Equations 4.54 and 4.56, we get

$$\frac{t_{f2}}{t_{f1}} = 2^{-2/3} = 0.63 \tag{4.57}$$

which means that we can reduce the time required to dissolve all of the material by 37% if we distribute the material in two particles instead of one.

Equation 4.47 was obtained for the case in which the liquid contains no A far from the sphere (boundary condition 4.44). If we rework the problem for the case in which the concentration of A in the liquid far from the sphere is $c_{A\infty}$, it can be shown that the dissolution flux is

$$N_{Ar}\big|_{r=R} = \frac{D_{AB}}{R}(c_{As} - c_{A\infty}) \tag{4.58}$$

We can see that this equation can be rewritten in terms of a mass transfer coefficient k_c as follows:

$$N_{Ar}\big|_{r=R} = k_c(c_{As} - c_{A\infty}) \tag{4.59}$$

where

$$k_c = \frac{D_{AB}}{R} \tag{4.60}$$

The mass transfer coefficient can be represented in dimensionless form in terms of the Sherwood number, which will gain importance in the treatment of convection problems in Chapter 6. The Sherwood number defined as follows:

$$Sh = \frac{k_c(2R)}{D_{AB}} \tag{4.61}$$

From Equation 4.60, we can see that

$$\mathrm{Sh} \equiv 2, \text{diffusion from a sphere} \tag{4.62}$$

4.3 TRANSIENT DIFFUSION

In this section, we will consider unsteady-state diffusion for cases in which the diffusing species is dilute and c and D_{AB} are uniform, so that the diffusion equation is applicable. In addition, we will treat cases without chemical reactions ($R_A = 0$).

4.3.1 Transient Mass Transfer in a Flat Membrane

Consider a flat solid membrane of thickness 2L (Figure 4.6) that initially has a uniform concentration of a species A and is brought into contact at $t = 0$ on both sides with a moving fluid that has a bulk concentration of A equal to $c_{A\infty}$. External mass transfer between the surface of the membrane and the fluid is governed by convection with a fluid-phase mass transfer coefficient k_c. We would like to find the concentration of A in the membrane as a function of position and time: $c_A = c_A(t, x)$.

The diffusion equation in rectangular coordinates (Equation C.11) in this case simplifies to the following partial differential equation (PDE),

$$\frac{\partial c_A}{\partial t} = D_{AB} \frac{\partial^2 c_A}{\partial x^2} \tag{4.63}$$

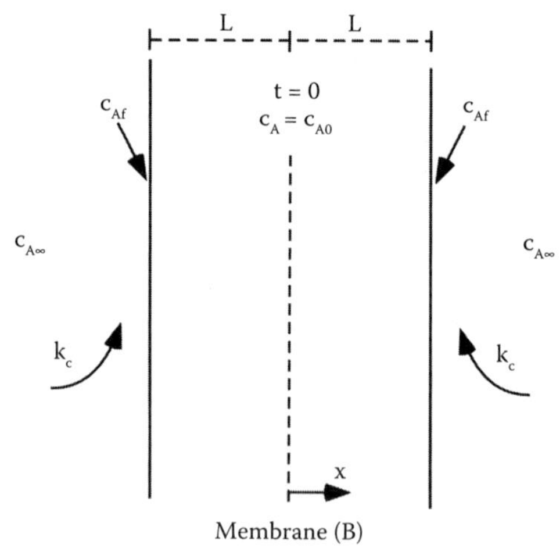

FIGURE 4.6 Transient mass transfer in a flat membrane surrounded by a moving fluid. The membrane is very large in the y and z directions. Species A diffuses through the solid membrane (B) and is transported from the membrane surface to the bulk fluid by convection.

The initial condition is

$$c_A = c_{A0}, t = 0 \tag{4.64}$$

A first boundary condition is due to the symmetry of the concentration profile:

$$\frac{\partial c_A}{\partial x} = 0, x = 0 \tag{4.65}$$

Because of symmetry, we need to solve the problem in one-half of the membrane only ($0 < x < L$). This condition is equivalent to stating that the molar flux of A on the plane $x = 0$ is zero, since

$$N_A = -D_{AB}\frac{\partial c_A}{\partial x} \tag{4.66}$$

A second boundary condition comes from a mole balance of A at the surface of the membrane ($x = L$): the molar flux of A that gets to the surface by diffusion must equal the molar flux of A transferred by convection to the bulk of the fluid:

$$-D_{AB}\left.\frac{\partial c_A}{\partial x}\right|_{x=L} = k_c(c_{Af} - c_{A\infty}) \tag{4.67}$$

The concentration of A in the fluid that is in contact with the surface of the membrane, c_{Af}, is at equilibrium with the concentration in the membrane. Here, we will assume a linear partitioning, so that

$$c_{Af} = K\left.c_A\right|_{x=L} \tag{4.68}$$

The boundary condition (4.67) can be then expressed as follows:

$$-D_{AB}\frac{\partial c_A}{\partial x} = k_c K\left(c_A - \frac{c_{A\infty}}{K}\right), x = L \tag{4.69}$$

The problem is fully formulated by the PDE 4.63 and the additional conditions 4.64, 4.65, and 4.66. The solution of this problem would yield $c_A = c_A(t, x)$. As formulated, the problem contains six parameters: D_{AB}, c_{A0}, L, k_c, K, and $c_{A\infty}$. That is, a change of any of these constants would change the concentration field. To minimize the number of parameters in the problem formulation, we will reformulate the problem in dimensionless form. We will select appropriate dimensionless variables to represent the independent variables (t and x) of the problem, as well as the dependent variable (c_A). The dimensionless formulation will also serve to establish appropriate scaling trends for the solution, as will be discussed later.

To cast the problem into dimensionless form, we define the following dimensionless variables:

$$\Theta = \frac{c_A - c_{A\infty}/K}{c_{A0} - c_{A\infty}/K} \tag{4.70}$$

$$\eta = \frac{x}{L} \tag{4.71}$$

$$\tau = \frac{D_{AB}t}{L^2} \tag{4.72}$$

Note that Θ represents a dimensionless concentration difference, η is dimensionless position, and τ is dimensionless time. Substituting these definitions into the differential Equation 4.63 and the initial and boundary conditions (4.64), (4.65), and (4.69) leads to, after manipulations,

$$\frac{\partial\Theta}{\partial\tau} = \frac{\partial^2\Theta}{\partial\eta^2} \tag{4.73}$$

$$\Theta = 1, \tau = 0 \tag{4.74}$$

$$\frac{\partial\Theta}{\partial\eta} = 0, \ \eta = 0 \tag{4.75}$$

$$\frac{\partial\Theta}{\partial\eta} = -\text{Bi}\Theta, \ \eta = 1 \tag{4.76}$$

where the mass transfer Biot number is a unitless parameter defined by

$$\text{Bi} = \frac{k_c KL}{D_{AB}} \tag{4.77}$$

The dimensionless formulation (Equations 4.73 through 4.76) has only one parameter (Bi), as opposed to the six parameters present in the original formulation. This means that the Biot number is the only physical parameter that affects the dimensionless concentration field, $\Theta(\tau, \eta)$. As we will discuss later, the Biot number can be interpreted as a comparative ratio between the external mass transfer process (represented by k_c in Equation 4.77) and internal diffusion (represented by D_{AB}).

Even though the dimensionless problem may seem mathematically complex, it has an analytical solution that can be found by using the method of separation of variables. The mathematical fundamentals of the method go beyond our scope (for details, see, e.g., Kreyszig, 1999). However, we will go through the solution process in detail to illustrate the process. We start by postulating a solution of the form

$$\Theta(\tau, \eta) = f(\eta)g(\tau) \tag{4.78}$$

After substitution in the differential equation and separation, we obtain

$$\frac{1}{g}\frac{dg}{d\tau} = \frac{1}{f}\frac{d^2f}{d\eta^2} = -\lambda^2 \tag{4.79}$$

The left-hand side of this equation only depends on τ while the right-hand side only depends on η, so both sides must be equal to a constant (that we have called $-\lambda^2$) whose sign has been chosen as negative to ensure an exponential decay with time (see below). The differential equation for g can be directly integrated

$$\frac{dg}{g} = -\lambda^2 d\tau \Rightarrow g = Ae^{-\lambda^2\tau} \tag{4.80}$$

while for f we get

$$\frac{d^2f}{d\eta^2} + \lambda^2 f = 0 \tag{4.81}$$

whose solution is

$$f = C_1\sin(\lambda\eta) + C_2\cos(\lambda\eta) \tag{4.82}$$

The boundary conditions (4.75) and (4.76) are separable:

$$g\frac{df}{d\eta} = 0 \Rightarrow \frac{df}{d\eta} = 0, \eta = 0 \tag{4.83}$$

$$g\frac{df}{d\eta} = -\mathrm{Bi}\,fg \Rightarrow \frac{df}{d\eta} = -\mathrm{Bi}\,f,\ \eta = 1 \tag{4.84}$$

The application of Equation 4.82 yields

$$C_1\lambda\cos(\lambda\eta) - C_2\lambda\sin(\lambda\eta)\Big|_{\eta=0} = 0 \Rightarrow C_1 = 0 \tag{4.85}$$

From Equation 4.84, we find

$$-C_2\lambda\sin\lambda = -C_2\mathrm{Bi}\cos\lambda \tag{4.86}$$

Since C_2 cannot be zero (this would lead to a trivial solution $\Theta \equiv 0$), we conclude that

$$\lambda\tan\lambda = \mathrm{Bi} \tag{4.87}$$

This equation has infinite solutions (see Figure 4.7). The infinite roots of this equation ($\lambda_1 < \lambda_2 < \lambda_3 \cdots$) are the eigenvalues of the **Sturm–Liouville problem** posed by Equations 4.81, 4.83, and 4.84. Equation 4.87 is the **eigenvalue condition**. This means that this problem has infinite solutions of the form 4.82 (with $C_1 = 0$), one for each eigenvalue:

$$f_n(\eta) = C_2\cos(\lambda_n\eta) \tag{4.88}$$

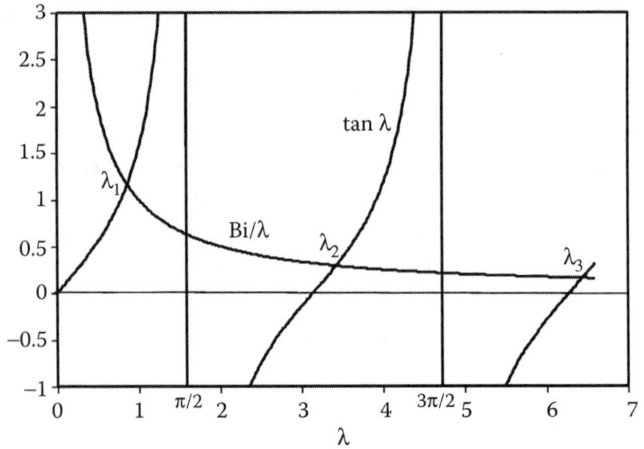

FIGURE 4.7 Graphical solution of Equation 4.87, interpreted as the intersection between the functions tan λ and Bi/λ. Values in the plot are for the case Bi = 1, but the trend is similar for any Bi. The first three roots are shown, but the periodicity of tan λ shows that infinite solutions exist: $0 < \lambda_1 \le (\pi/2) \le \lambda_2 \le (3\pi/2) \le \lambda_3 \le (5\pi/2)\cdots$

Similarly, there will be infinite functions g(τ) from Equation 4.80,

$$g_n(\tau) = Ae^{-\lambda_n^2\tau} \tag{4.89}$$

Up to this point, the only original equation in the problem formulation that we have not used is the initial condition (Equation 4.74). That condition, being nonhomogeneous is nonseparable; that is, it does not admit a solution of the form given by Equation 4.78. The theory of separation of variables (Kreyszig, 1999) indicates that the general solution of the problem up to this point is the linear combination of all the solutions given by Equation 4.78:

$$\Theta(\tau,\eta) = \sum_{n=1}^{\infty} a_n \cos(\lambda_n\eta)e^{-\lambda_n^2\tau} \tag{4.90}$$

The sum on the right-hand side of this equation is a **Fourier series**. At this point, the series coefficients, a_n, are not identified. We will determine these coefficients by applying the initial condition: evaluating the solution at the initial condition (4.74) yields

$$1 = \sum_{n=1}^{\infty} a_n \cos(\lambda_n\eta) \tag{4.91}$$

To find each coefficient, we will use the orthogonality condition satisfied by solutions of a Sturm–Liouville problem (Kreyszig, 1999), which in this case states that, for any pair of different eigenvalues λ_n, λ_m,

$$\int_0^1 \cos(\lambda_n\eta)\cos(\lambda_m\eta)\,d\eta = 0, \text{ for } n \neq m \tag{4.92}$$

We now multiply both sides of Equation 4.91 by $\cos(\lambda_m\eta)$ and integrate from $\eta = 0$ to $\eta = 1$. Equation 4.92 implies that all the terms in the sum on the right-hand side of the resulting equation will be zero, except the term for which n = m, so that we find

$$\int_0^1 \cos(\lambda_m\eta)\,d\eta = a_m \int_0^1 \cos^2(\lambda_m\eta)\,d\eta \tag{4.93}$$

Evaluating the integrals leads to

$$a_m = \frac{2\sin\lambda_m}{\lambda_m + \sin\lambda_m\cos\lambda_m} \tag{4.94}$$

Letting m = n and substituting into Equation 4.90, we find the final solution:

$$\Theta(\tau,\eta) = \sum_{n=1}^{\infty} \frac{2\sin\lambda_n}{\lambda_n + \sin\lambda_n\cos\lambda_n}\cos(\lambda_n\eta)e^{-\lambda_n^2\tau} \tag{4.95}$$

The Fourier series in this equation is uniformly convergent, which means that the value of each term decreases as more terms are evaluated. The factor that controls its convergence is the exponential term, which rapidly becomes small as n increases (recall that λ_n increases with n). Note that, the larger the value of τ, the faster the decay of the exponential with increasing n. In fact, detailed calculations show that, if $Bi \geq 0.01$, the first term of the series is enough to give the value of Θ with accuracy of 0.1% or better, as long as $\tau \geq 0.1$. Under the conditions stated, the approximate solution would be

$$\Theta(\tau,\eta) \cong \frac{2\sin\lambda_1}{\lambda_1 + \sin\lambda_1\cos\lambda_1}\cos(\lambda_1\eta)e^{-\lambda_1^2\tau}, \; Bi \geq 0.01, \; \tau \geq 0.1 \tag{4.96}$$

The values of λ_1 (first root of the eigenvalue condition 4.87) are shown in Figure 4.8 as a function of the Biot number. Note that $\lambda_1 \to 0$ as $Bi \to 0$, and that $\lambda_1 \to \pi/2$ (1.5708) as $Bi \to \infty$.

To determine the molar flux of A at the surface of the membrane, we average the differential Equation 4.63 over the membrane volume as follows:

$$\frac{1}{L}\int_0^L \frac{\partial c_A}{\partial t}dx = \frac{1}{L}\int_0^L D_{AB}\frac{\partial^2 c_A}{\partial x^2}dx \tag{4.97}$$

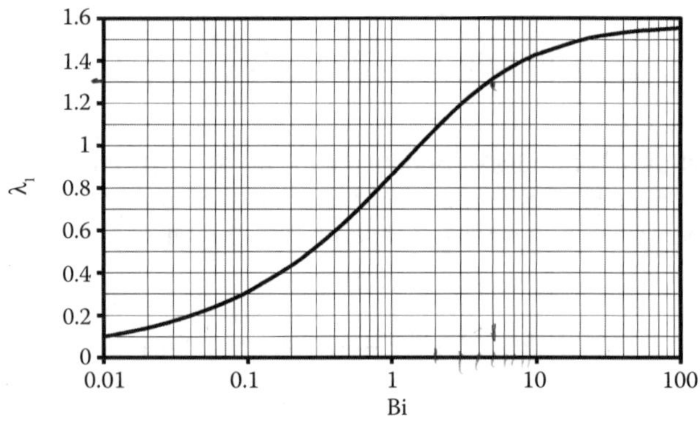

FIGURE 4.8 First root of Equation 4.87 as a function of the Biot number.

Realizing that

$$<c_A> = \frac{1}{L}\int_0^L c_A \, dx \tag{4.98}$$

is the average molar concentration of A in the membrane, Equation 4.97 becomes, after integrating the right-hand side

$$\frac{d <c_A>}{dt} = \frac{D_{AB}}{L}\frac{\partial c_A}{\partial x}\bigg|_0^L \tag{4.99}$$

which leads to, after use of Equation 4.66,

$$N_{Ax}\big|_{x=L} = -L\frac{d <c_A>}{dt} \tag{4.100}$$

From Equation 4.70, we see that

$$<c_A> = (c_{A0} - c_{A\infty}/K) <\Theta> + c_{A\infty}/K \tag{4.101}$$

The average dimensionless concentration can be found by using the dimensionless version of Equation 4.98:

$$<\Theta> = \int_0^1 \Theta \, d\eta \tag{4.102}$$

Substituting the solution (Equation 4.95) into this equation and integrating leads to

$$<\Theta> = \sum_{n=1}^{\infty} \frac{2\sin^2\lambda_n}{(\lambda_n + \sin\lambda_n \cos\lambda_n)\lambda_n} e^{-\lambda_n^2\tau} \tag{4.103}$$

The flux can now be calculated combining Equations 4.100, 4.101, and 4.103, which yields

$$N_{Ax}\Big|_{x=L} = \frac{D_{AB}}{L}(c_{A0} - c_{A\infty}/K)\sum_{n=1}^{\infty} \frac{2\lambda_n \sin^2\lambda_n}{\lambda_n + \sin\lambda_n \cos\lambda_n} e^{-\lambda_n^2\tau} \tag{4.104}$$

The solution presented corresponds to a flat plate exposed to fluid on both sides (Figure 4.6). However, we have taken advantage of the symmetry of the temperature profiles to find the solution in the half plate $0 < x < L$. Note that the symmetry boundary condition at the centerline (Equation 4.65) also implies that $N_{Ax} = 0$ at $x = 0$. This means that the solution found is also applicable to a plate of thickness L exposed to fluid on one side ($x = L$) and in contact with a surface impermeable to A at $x = 0$.

An interesting limiting case of the solution presented above occurs when convective mass transfer to the surroundings is much faster than diffusion in the body. In this case, Bi » 1, which implies that the concentration of A in the solid at the surface is equal to the concentration at equilibrium with the fluid bulk; that is, $c_A\big|_{x=L} = c_{A\infty}/K$. The solution is given by the limit of the above equations when $Bi \to \infty$. The eigenvalue condition (Equation 4.87) yields $\tan\lambda \to \infty$, which implies that the eigenvalues are

$$\lambda_n = \frac{(2n-1)\pi}{2}, \; n = 1, 2, \ldots, (Bi \to \infty) \tag{4.105}$$

The dimensionless concentration (Equation 4.95) then becomes

$$\Theta(\tau, \eta) = \frac{4}{\pi}\sum_{n=1}^{\infty} \frac{(-1)^{n-1}}{(2n-1)} \cos\left[\frac{(2n-1)\pi\eta}{2}\right] e^{-((2n-1)^2\pi^2\tau/4)} \tag{4.106}$$

and the average concentration (Equation 4.103) is

$$<\Theta> = \frac{8}{\pi^2}\sum_{n=1}^{\infty} \frac{1}{(2n-1)^2} e^{-((2n-1)^2\pi^2\tau/4)} \tag{4.107}$$

This is the same solution that would be obtained using separation of variables replacing the convective boundary condition (4.76) by

$$\Theta = 0, \eta = 1 \tag{4.108}$$

The dimensionless concentration profile given by Equation 4.106 is plotted in Figure 4.9 for various values of dimensionless time. It is interesting to see that, for short times ($\tau \leq 0.04$), the center of the plate remains at the initial concentration ($\Theta = 1$) since the perturbation imposed by exposing the surface of the plate to the fluid does not have time to reach the center. Note also that a dimensionless time $\tau = 1$ can be considered a long time in this process, since the plate temperature is close to its steady-state value ($\Theta = 0$).

EXAMPLE 4.1 LEACHING OF A FLAME RETARDANT FROM A CARPET

Flame retardants are compounds added to common products to prevent them from catching fire, which makes them safe for household applications. However, there has been concern recently on possible health effects on people who come in contact with products containing these compounds (e.g., Whitacre, 2013). An experiment has been designed to measure the diffusivity of a particular flame retardant (A) through carpet fabric (B). A flat carpet, measuring 2 m × 2 m with thickness 3 cm, is placed on the floor and subjected to a very fast flow of pure air over its top surface. The carpet initially contains 120 g of flame retardant. Measures conducted show that 50 g of flame retardant leached out from the top surface of the carpet after 90 days.

1. Determine D_{AB} with the information given.
2. Under normal conditions, when air is not rapidly flowing over the carpet, convection on the carpet surface yields a mass transfer Biot number of Bi = 0.1. Calculate how long it will take for the original carpet to lose 50 g of

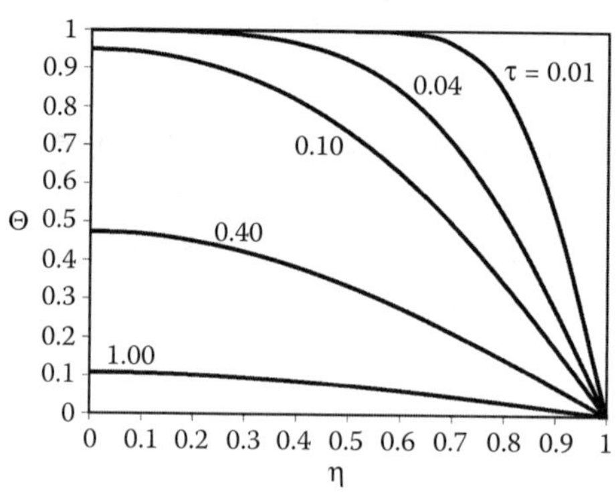

FIGURE 4.9 Dimensionless temperature profile for transient conduction in a flat plate with Bi → ∞.

flame retardant under these conditions. Consider that the concentration of flame retardant in the air is always negligibly low.

3. The information given allows us to calculate the dimensionless average concentration of flame retardant in the carpet at t = 90 days. To do this, we will use Equation 4.70. In this case, $c_{A\infty} = 0$, so that

$$<\Theta> = \frac{<c_A>}{c_{A0}} = \frac{m_A}{m_{A0}} = \frac{70}{120} = 0.583$$

Note that we are using mass concentrations instead of molar concentrations in these calculations.

The carpet is a flat plate with no mass transfer at the bottom surface, so we can use the solution for a plate with thickness 2L transferring mass on both sides if we let L = 3 cm. Since the air flow is very fast, we can state that Bi → ∞, and we will assume that dimensionless time is high enough that the first term in the Fourier series solution is a good approximation to the solution, so that Equation 4.96 applies with $\lambda_1 = \pi/2$, which yields

$$<\Theta> = \frac{8}{\pi^2} e^{(\pi^2/4)\tau} \Rightarrow \tau = -\frac{4}{\pi^2} \ln\left(\frac{\pi^2}{8} <\Theta>\right)$$

This gives $\tau = 0.1336$, which confirms our assumption that $\tau \geq 0.1$. From the definition of dimensionless time, we get

$$D_{AB} = \frac{\tau L^2}{t} \Rightarrow D_{AB} = 1.55 \times 10^{-7}\ \text{cm}^2/\text{s}$$

In this case, for Bi = 0.1, the first eigenvalue is calculated from Equation 4.87 to be $\lambda_1 = 0.311$. Now, since $<\Theta> = 0.583$, we can calculate τ directly

$$\tau = -\frac{1}{\lambda_1^2} \ln\left(\frac{<\Theta>\ (\lambda_1 + \sin\lambda_1 \cos\lambda_1)\lambda_1}{2\sin^2\lambda_1}\right) = 5.61$$

which implies

$$t = \frac{\tau L^2}{D_{AB}} = 3.26 \times 10^8\ \text{s}$$

4.3.2 Transient Mass Transfer in Long Cylinders and Spheres

The methodology applied to the flat plate problem can be extended to other simple geometries. The method of separation of variables yields analytical solutions of the diffusion equation for infinitely long cylinders and spheres.

For an infinitely long cylinder of radius R containing species A at a uniform initial concentration c_{A0} that is being exposed to a fluid at a concentration $c_{A\infty}$ far from the cylinder, the concentration field inside the cylinder is a function of time and radial position. The problem is formulated using the dimensionless concentration given by Equation 4.70 and dimensionless time and radial position defined by

$$\tau = \frac{D_{AB}t}{R^2} \tag{4.109}$$

$$\xi = \frac{r}{R} \tag{4.110}$$

The Biot number in this case is given by

$$Bi = \frac{k_c KR}{D_{AB}} \tag{4.111}$$

It can be shown that the dimensionless diffusion equation in cylindrical coordinates simplifies to

$$\frac{\partial \Theta}{\partial \tau} = \frac{1}{\xi}\frac{\partial}{\partial \xi}\left(\xi \frac{\partial \Theta}{\partial \xi}\right) \tag{4.112}$$

and the initial and boundary conditions are

$$\Theta = 1, \tau = 0 \tag{4.113}$$

$$\frac{\partial \Theta}{\partial \xi} = 0, \xi = 0 \tag{4.114}$$

$$\frac{\partial \Theta}{\partial \xi} = -Bi\Theta, \xi = 1 \tag{4.115}$$

In this case, the solution is expressed in terms of Bessel functions (Appendix D provides a brief review of Bessel functions). The eigenvalue condition is

$$\lambda_n \frac{J_1(\lambda_n)}{J_0(\lambda_n)} = Bi \tag{4.116}$$

which, as was the case for the flat plate problem, has infinite solutions. The first root of this equation is presented in Figure 4.10.

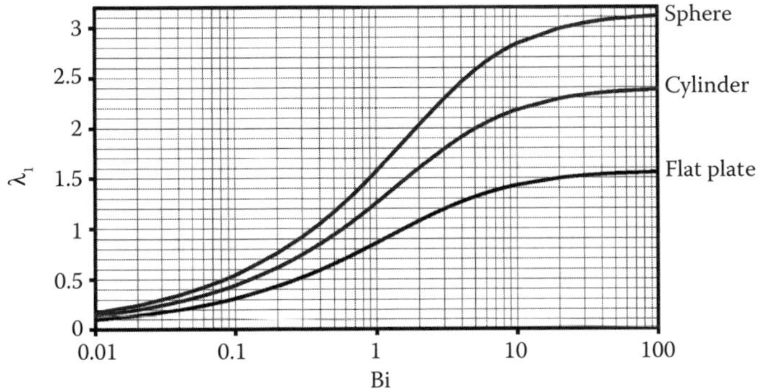

FIGURE 4.10 First eigenvalue as a function of the Biot number for flat plate (same as in Figure 4.8), infinitely long cylinder (first root of Equation 4.116), and sphere (first root of Equation 4.124).

The solution of the problem (Equations 4.112 through 4.115) is

$$\Theta = \sum_{n=1}^{\infty} \frac{2}{\lambda_n} \frac{J_1(\lambda_n)}{J_0^2(\lambda_n) + J_1^2(\lambda_n)} J_0(\lambda_n \xi) e^{-\lambda_n^2 \tau} \tag{4.117}$$

The first term of the series is a good approximation for Bi ≥ 0.01 and $\tau \geq 0.1$. The first coefficient of the series is

$$a_1 = \frac{2}{\lambda_1} \frac{J_1(\lambda_1)}{J_0^2(\lambda_1) + J_1^2(\lambda_1)} \tag{4.118}$$

and is shown as a function of Biot number in Figure 4.11.

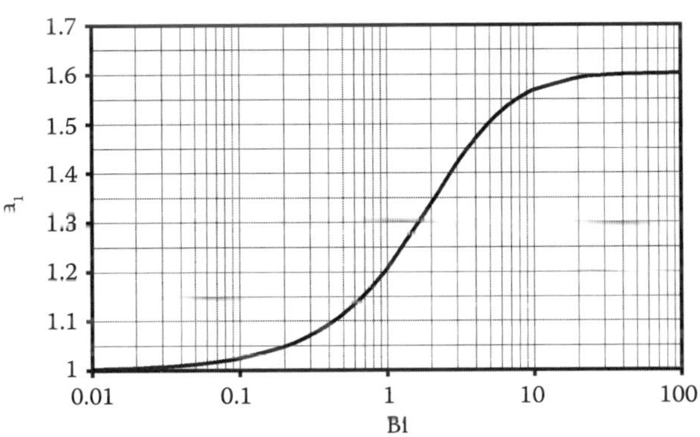

FIGURE 4.11 First coefficient of the Fourier series for the cylinder, Equation 4.117.

The average dimensionless concentration in this case is given by

$$<\Theta> = \sum_{=1}^{\infty} \frac{4}{\lambda_n^2} \frac{J_1^2(\lambda_n)}{J_0^2(\lambda_n) + J_1^2(\lambda_n)} e^{-\lambda_n^2 \tau} \tag{4.119}$$

For the case of a sphere at a uniform initial concentration c_{A0} that is being exposed to a fluid at a concentration $c_{A\infty}$ far from the sphere, the concentration field inside the sphere will be a function of time and radial position. The formulation of this problem is

$$\frac{\partial \Theta}{\partial \tau} = \frac{1}{\xi^2} \frac{\partial}{\partial \xi}\left(\xi^2 \frac{\partial \Theta}{\partial \xi}\right) \tag{4.120}$$

(diffusion equation in spherical coordinates), with initial and boundary conditions

$$\Theta = 1, \tau = 0 \tag{4.121}$$

$$\frac{\partial \Theta}{\partial \xi} = 0, \xi = 0 \tag{4.122}$$

$$\frac{\partial \Theta}{\partial \xi} = -\mathrm{Bi}\Theta, \xi = 1 \tag{4.123}$$

The dimensionless variables follow the same definitions as in the cylinder case, using R as the sphere radius.

The eigenvalue condition for this problem is

$$1 - \lambda_n \cot \lambda_n = \mathrm{Bi} \tag{4.124}$$

The first eigenvalue is plotted in Figure 4.10. The solution is

$$\Theta = \sum_{n=1}^{\infty} \frac{4}{\lambda_n} \frac{\sin\lambda_n - \lambda_n \cos\lambda_n}{2\lambda_n - \sin 2\lambda_n} \frac{\sin(\lambda_n \xi)}{\xi} e^{-\lambda_n^2 \tau} \tag{4.125}$$

The dimensionless concentration at the center of the sphere can be found using the limit (which can be found using L'Hopital's rule)

$$\lim_{\xi \to 0} \frac{\sin(\lambda_n \xi)}{\xi} = \lambda_n \tag{4.126}$$

and the average concentration is given by

$$<\Theta> = \sum_{n=1}^{\infty} \frac{12}{\lambda_n^3} \frac{(\sin\lambda_n - \lambda_n \cos\lambda_n)^2}{2\lambda_n - \sin 2\lambda_n} e^{-\lambda_n^2 \tau} \tag{4.127}$$

EXAMPLE 4.2 LOADING OF FERTILIZER SPHERES

Solid polysaccharide (B) spheres (1 cm radius) loaded with a fertilizer (A) will be used in controlled-release applications. The spheres have initially no fertilizer and will be loaded in a pressurized tank where high-pressure air containing 0.01 mol/cm³ of fertilizer will flow continuously over the spheres. At equilibrium, the distribution of fertilizer between air and the solid is given by

$$c_{A,gas} = 0.1 c_{A,solid} \tag{4.128}$$

The mass transfer coefficient for the process is $k_c = 0.01$ cm/s. The diffusivity of the fertilizer in the polysaccharide spheres is $D_{AB} = 1 \times 10^{-4}$ cm²/s. We wish to answer the following questions:

1. What is the maximum concentration of the fertilizer that can be achieved in the solid spheres by this process?
2. How long will it take to load the spheres with an average concentration of 0.095 mol/cm³ of the fertilizer?
3. The loaded spheres are isolated from the surroundings in a storage tank for a few days. Afterwards, they are exposed for 2 h to air with 0.02 mol/cm³ of the fertilizer. What is the final concentration of the fertilizer **at the center** of the spheres?

1. The maximum concentration that the solid can have occurs when it achieves equilibrium with the gas; that is,

$$c_{A\max,solid} = \frac{c_{A\infty}}{0.1} = 0.1 \text{ mol/cm}^3 \tag{4.129}$$

2. The average dimensionless concentration for the sphere is given by Equation 4.127. If we assume $\tau > 0.1$, the first term is accurate enough,

$$<\Theta> \approx \frac{12}{\lambda_1^3} \frac{(\sin\lambda_1 - \lambda_1 \cos\lambda_1)^2}{2\lambda_1 - \sin 2\lambda_1} e^{-\lambda_1^2 \tau} \tag{4.130}$$

where the first eigenvalue is calculated from

$$1 - \lambda_n \cot\lambda_n = Bi \tag{4.131}$$

In this case

$$Bi = \frac{k_c K R}{D_{AB}} = \frac{0.01 \times 0.1 \times 1}{10^{-4}} = 10 \tag{4.132}$$

We find $\lambda_1 = 2.84$. At the end of the process, we know that

$$<\Theta> = \frac{<c_A> - c_{A\infty}/K}{c_{A0} - c_{A\infty}/K} = \frac{0.095 - 0.10}{-0.10} = 0.05 \tag{4.133}$$

Therefore,

$$\tau = -\frac{1}{\lambda_1^2}\ln\left[\frac{<\Theta>\lambda_1^3(2\lambda_1 - \sin 2\lambda_1)}{12(\sin\lambda_1 - \lambda_1\cos\lambda_1)^2}\right] = 0.3384 \tag{4.134}$$

which yields

$$t = \frac{R^2\tau}{D_{AB}} = 3384 \text{ s} = 56.4\text{min} \tag{4.135}$$

3. The spheres will have an average concentration of 0.095 mol/cm³. This average concentration will become uniform during isolation. For the new process:

$$c_{A0} = 0.095 \text{ mol/cm}^3 \tag{4.136}$$

$$c_{A\infty} = 0.02 \text{ mol/cm}^3 \tag{4.137}$$

$$\tau = \frac{tD_{AB}}{R^2} = 0.72 \tag{4.138}$$

The solution for the dimensionless concentration is (Equation 4.125) (only first term)

$$\Theta = \frac{4}{\lambda_1}\frac{\sin\lambda_1 - \lambda_1\cos\lambda_1}{2\lambda_1 - \sin 2\lambda_1}\frac{\sin(\lambda_1\xi)}{\xi}e^{-\lambda_1^2\tau} \tag{4.139}$$

At the center ($\xi = 0$), we have

$$\Theta = 4\frac{\sin\lambda_1 - \lambda_1\cos\lambda_1}{2\lambda_1 - \sin 2\lambda_1}e^{-\lambda_1^2\tau} \tag{4.140}$$

Since the eigenvalue is still the same, we find $\Theta = 0.005874$, which yields

$$0.005874 = \frac{c_A - c_{A\infty}/K}{c_{A0} - c_{A\infty}/K} \Rightarrow c_A = 0.1994 \text{mol/cm}^3 \tag{4.141}$$

4.3.3 Transient Leaching from a Semi-Infinite Solid

Consider that a solid containing a uniform concentration of a soluble salt c_{A0} is put in contact at t = 0 with a large volume of liquid with a uniform, constant and relatively low salt concentration c_{Al}. Here, we will treat the case in which a flat surface of a large solid body is contacted with the liquid, so that the solid–liquid interface can be considered of infinite surface area, and the solid phase can be considered of semi-infinite thickness, as shown in Figure 4.12.

The symmetry of the geometry leads to conclusion that the concentration in the solid is only a function of t and x: $c_A = c_A(t, x)$. The mass transfer of solute in the solid is governed by the diffusion equation:

$$\frac{\partial c_A}{\partial t} = D_A \frac{\partial^2 c_A}{\partial x^2} \tag{4.142}$$

In this case, D_A is the diffusivity of salt in the solid. The initial condition is

$$c_A = c_{A0},\ t = 0 \tag{4.143}$$

Fast convective mass transfer in the liquid (perfect mixing) combined with the assumption of linear partitioning of the salt between liquid and solid phases leads to

$$c_A = Kc_{Al},\ x = 0 \tag{4.144}$$

Here, we will consider that the concentration of A in the liquid is low enough that we can state that $c_{Al} \approx 0$. The fact that the solids extend indefinitely in the x direction means that there will be a region far from the interface that will remain at the initial concentration:

$$c_A = c_{A0}, x \to \infty \tag{4.145}$$

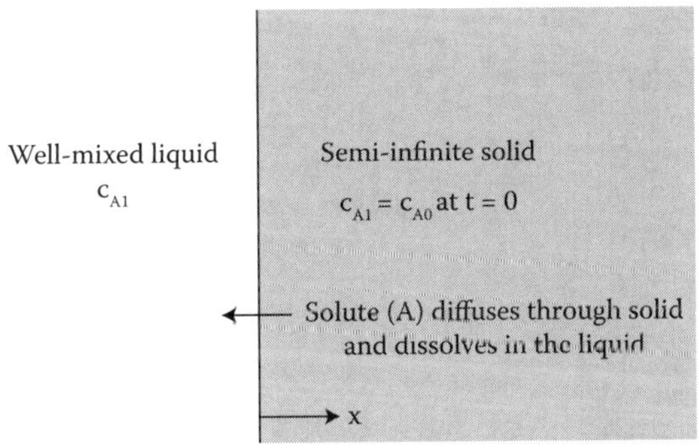

FIGURE 4.12 Leaching from a semi-infinite solid: the solid phase occupies the space $0 < x < \infty$, and contains initially a concentration of solute c_{A0}. The liquid that contacts the solid is perfectly mixed and its volume so large, that the concentration of salt in the liquid can be assumed to be constant, c_{Al}, during the leaching process.

To solve this problem, we will use the method of combination of variables (similarity transformation). Mathematically, the similarity transformation is based on postulating that the concentration field can be expressed in terms of a single independent variable, $c_A = c_A(\eta)$, that is a combination of the independent variables t and x. Here, we will postulate that the similarity variable has the form

$$\eta = \frac{x}{\delta(t)} \tag{4.146}$$

This implies that the concentration profile will look the same at all times (i.e., all profiles are "similar"), with the only difference that it will be stretched by the magnitude of the length scale δ(t). To verify if the postulate made is true, the differential equation, as well as the initial and boundary conditions, must be consistent with it. The boundary condition 4.144 with $c_{AI} \approx 0$ will be applied at $\eta = 0$:

$$c_A = 0, \eta = 0 \tag{4.147}$$

Since η is potentially the only independent variable, this means that the conditions 4.143 and 4.145 must collapse to a single condition at a specific value of η, since both are identical conditions on c_A. By inspection of Equation 4.146, this implies, necessarily, that

$$\delta(0) = 0 \tag{4.148}$$

in which case Equations 4.143 and 4.145 both lead to

$$c_A = c_{A0}, \eta \to \infty \tag{4.149}$$

Next, we transform the PDE by assuming that $c_A = c_A(\eta)$. Applying the chain rule and using Equation 4.146, we get

$$\frac{\partial c_A}{\partial t} = \frac{dc_A}{d\eta}\frac{\partial \eta}{\partial t} = -\frac{dc_A}{d\eta}\frac{x}{\delta^2}\frac{d\delta}{dt} = -\eta\frac{dc_A}{d\eta}\frac{1}{\delta}\frac{d\delta}{dt} \tag{4.150}$$

$$\frac{\partial c_A}{\partial x} = \frac{dc_A}{d\eta}\frac{\partial \eta}{\partial x} = \frac{dc_A}{d\eta}\frac{1}{\delta} \tag{4.151}$$

$$\frac{\partial^2 c_A}{\partial x^2} = \frac{1}{\delta}\frac{\partial}{\partial x}\left(\frac{dc_A}{d\eta}\right) = \frac{1}{\delta^2}\frac{d^2 c_A}{d\eta^2} \tag{4.152}$$

Substituting Equations 4.150 and 4.152 into the differential equation, Equation 4.142, leads to

$$-\eta\frac{dc_A}{d\eta}\frac{1}{\delta}\frac{d\delta}{dt} = D_A\frac{1}{\delta^2}\frac{d^2 c_A}{d\eta^2} \tag{4.153}$$

Letting $c_A' = dc/d\eta$, this equation can be separated as follows:

$$-\frac{c_A''}{\eta c_A'} = \frac{\delta}{D_A}\frac{d\delta}{dt} \tag{4.154}$$

The only way that this equation can be considered an ODE for $c_A(\eta)$ is if the right-hand side is a constant:

$$-\frac{c_A''}{\eta c_A'} = \frac{\delta}{D_A}\frac{d\delta}{dt} = \alpha \tag{4.155}$$

where the constant α can have any arbitrary value. This allows us to determine the function $\delta(t)$, by separating and integrating, and using Equation 4.148 we find

$$\int_0^\delta \delta d\delta = \alpha D_A \int_0^t dt \tag{4.156}$$

which leads to

$$\delta(t) = \sqrt{2\alpha D_A t} \tag{4.157}$$

Since the choice of α is arbitrary, we select $\alpha = 1$, which leads to the similarity variable

$$\eta = \frac{x}{\sqrt{2D_A t}} \tag{4.158}$$

and the concentration must satisfy Equation 4.155:

$$c_A'' = -\eta c_A' \tag{4.159}$$

subject to boundary conditions 4.147 and 4.149. To solve the ODE 4.159, we start with the transformation

$$u = \frac{dc_A}{d\eta} \tag{4.160}$$

which turns Equation 4.159 into

$$\frac{du}{d\eta} = -\eta u \tag{4.161}$$

Separating and integrating yields

$$u = Be^{-\eta^2/2} \tag{4.162}$$

where B is an integration constant. This leads to

$$\frac{dc_A}{d\eta} = Be^{-\eta^2/2} \tag{4.163}$$

Integrating and using condition 4.147 leads to

$$c_A = B\int_0^{\eta} e^{-\eta^2/2} d\eta \tag{4.164}$$

At this point, we make use of the definition of the error function

$$\mathrm{erf}(y) = \frac{2}{\sqrt{\pi}}\int_0^{y} e^{-y^2}\, dy \tag{4.165}$$

Figure 4.13 shows a plot of the error function. Using the transformation $y = \eta/\sqrt{2}$, we get

$$\int_0^{\eta} e^{-\eta^2/2}\, d\eta = \frac{\sqrt{2\pi}}{2}\mathrm{erf}\left(\frac{\eta}{\sqrt{2}}\right) \tag{4.166}$$

Therefore, Equation 4.164 becomes

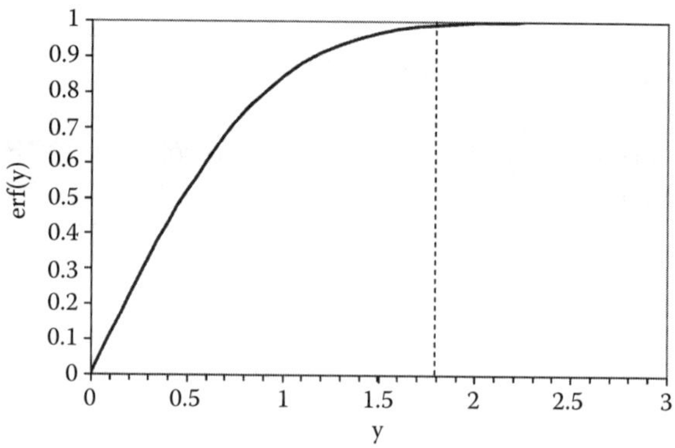

FIGURE 4.13 The error function. The dashed line represents the value of y for which erf(y) = 0.99.

$$c_A = B\sqrt{\frac{\pi}{2}}\,\mathrm{erf}\left(\frac{\eta}{\sqrt{2}}\right) \tag{4.167}$$

To find the integration constant, we apply the boundary condition 4.149. Knowing that erf(∞) = 1 (Figure 4.13) leads to the solution

$$c_A = c_{A0}\,\mathrm{erf}\left(\frac{x}{\sqrt{4D_At}}\right) \tag{4.168}$$

Now, we would like to calculate the rate at which the solute leaches from the solid. The leaching rate will be represented by the flux of A at the solid–liquid interface (Figure 4.12), $-N_{Ax}\big|_{x=0}$. To evaluate this flux, we use

$$N_{Ax} = -D_A\frac{\partial c_A}{\partial x} \tag{4.169}$$

Substituting Equation 4.168 into Equation 4.169, and taking the derivative of the error function leads to, after manipulations,

$$N_{Ax} = -c_{A0}\sqrt{\frac{D_A}{\pi t}}\,e^{-\eta^2/2} \tag{4.170}$$

At $x = 0$ ($\eta = 0$), we find

$$-N_{Ax}\big|_{x=0} = c_{A0}\sqrt{\frac{D_A}{\pi t}} \tag{4.171}$$

An interesting aspect of the solution for the concentration in the solid (Equation 4.168) is the fact that, at a given time, even though c_{A0} is reached asymptotically as $x \to \infty$, it is possible to define a position x at which the concentration of solute has reached c_{A0} approximately. For example, if we let

$$x = \beta, \text{ for } c_A = 0.99c_{A0} \tag{4.172}$$

we can state that at position β the concentration has reached the asymptotic value for all practical purposes. From Equation 4.168, this means that

$$0.99 = \mathrm{erf}\left(\frac{\beta}{\sqrt{4D_At}}\right) \tag{4.173}$$

The point at which the error function reaches a value of 0.99 occurs when the argument is 1.80 (Figure 4.13). Therefore, we find

$$\beta = 3.6\sqrt{D_At} \tag{4.174}$$

This equation allows us to calculate the distance that the concentration profile has penetrated into the solid phase at a given time. In fact, this equation can be used as a criterion to determine when a solid can be considered semi-infinite. For example, if we look at the solution obtained for a flat membrane in Section 4.3.1, at times short enough that the concentration profile has not penetrated the thickness of the half plate (L), the membrane can be considered a semi-infinite solid and Equation 4.168 would yield a good approximation for the concentration profile (note that the coordinate x in this equation is defined differently from the coordinate x used in the solution in Section 4.3.1). In other words, based on the criterion established in Equation 4.172, the membrane can be approximately considered a semi-infinite solid when

$$3.6\sqrt{D_{AB}t} < L \tag{4.175}$$

which means

$$\tau = \frac{D_{AB}t}{L^2} < 0.077 \tag{4.176}$$

This is precisely a range of times for which several terms in the Fourier series are required to give an accurate solution of the concentration profile for the flat plate. We see that using the Fourier series would be unnecessary, since the solution is well represented by Equation 4.168.

PROBLEMS

4.1 In controlled release systems, pills are designed to release a constant molar rate of a specific drug (A) to the patient. A typical controlled release pill is a hollow sphere filled with a saturated solution of the drug in water, which contains small crystals of the solid drug (see figure). The purpose of the crystals is to keep the solution inside the pill saturated at all times. The drug diffuses through a solid membrane (B), and is carried away by the fluid flowing around the pill. To study this process in the laboratory, a pill was suspended in a constant flow of water intended to simulate the external mass transfer process that takes place inside the patient's digestive tract. The following information is available: the concentration of the drug in the membrane that is at equilibrium with a solution with concentration c_A is given by $c_{Am} = Kc_A$, where $K = 0.2$. The diffusivity of A in the membrane is $D_{AB} = 1.50 \times 10^{-6}$ cm²/s. Convection around the pill is characterized by a mass transfer coefficient $k_c = 1.9 \times 10^{-3}$ cm/s. The drug A is dilute in water and within the membrane. The inner and outer radii of the membrane are $R_1 = 3$ mm and $R_2 = 4$ mm, respectively. The solubility of the drug in water is 2×10^{-5} mol/cm³. Determine:

a. The steady-state concentration profile of A in the membrane.
b. The rate of release of A from the pill into the flowing water (moles of A released per unit time).

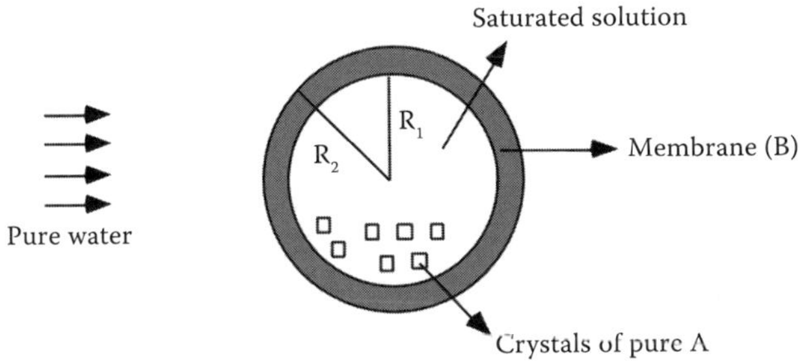

4.2 In a dialysis process, blood flows through the core of a cylindrical membrane. The toxins (represented by a single component A) in the blood are transported by diffusion through the membrane (B) to an outside solution (dialysis liquid) that is moving very fast and is kept free of toxins at all times (see figure). Over time, a layer of proteins deposits on the surface of the membrane forming a gel phase. The gel represents a diffusional resistance for the toxin, thus decreasing the efficiency of the process.

Find an expression to calculate the moles of toxin per unit time (W_A) that are transported through the membrane at steady state. Assume that convective mass transfer between the blood and the gel layer is characterized by a mass transfer coefficient k_c. Consider that the following parameters and facts are known: diffusivity of toxin in the membrane, D_{AB}; diffusivity of toxin in the gel layer, D_{AG}; inner membrane radius, R; length of membrane, L; membrane thickness, δ; gel layer thickness, d; concentration of toxin in the blood, c_{Ai}. At equilibrium, the ratio between toxin concentrations in the membrane and the gel layer is

$$K = \left(\frac{c_{A,\text{membrane}}}{c_{A,\text{gel}}}\right)_{\text{equil}}$$

Also at equilibrium, the protein gel and the blood have the same concentration of toxin.

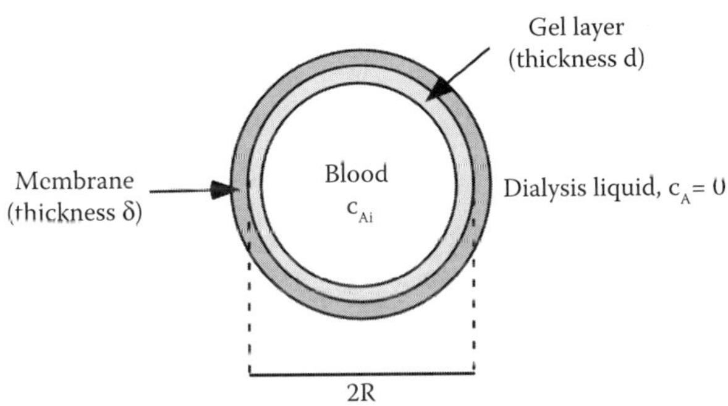

4.3 A spherical water droplet with an initial diameter of 2.5 mm is suspended on a thin wire in a large volume of stationary air at 30°C and 1 atm. The air has a relative humidity of 5%. Estimate the time required for complete evaporation of the water droplet assuming that its temperature does not change during evaporation. The diffusivity of water vapor in air at 30°C is 0.27 cm²/s. Recall that relative humidity is defined by

$$\phi(\%) = \frac{P_w}{P_v} \times 100$$

where P_w is the partial pressure of water vapor in the air and P_v is the vapor (saturation) pressure of water vapor in the air at the mixture temperature.

4.4 The amount of oxygen transferred to the anterior chamber of the eye through the cornea varies considerably depending on whether a contact lens is worn or not. Treat the eye as a spherical system, and assume that mass transfer between the cornea and the anterior chamber, and between the surroundings and the eye occurs by convection, characterized by mass transfer coefficients k_o and k_i (see figure). The system is at steady state and the cornea and the lens cover one-third of the spherical surface area. Determine the mole transfer rate of oxygen (mol/s) between the surroundings and the eye's anterior chamber with and without the contact lens in place. The partition coefficient of oxygen between the air and the cornea is

$$K_P = \frac{c_{air}}{c_{cornea}} = 10.5$$

Assume that this is also the partition coefficient of oxygen between air and the contact lens.

Data:

Dimensions:

$r_1 = 10.2$ mm, $r_2 = 12.7$ mm, $r_3 = 16.5$ mm

Oxygen concentrations:

$c_{\infty I} \approx 0$, $c_{\infty o} = 0.01$ mol/L

Mass transfer coefficients:

$k_i = 0.3$ cm/s, $k_o = 1$ cm/s

Diffusivities of oxygen:

$D_1 = 1.2 \times 10^{-5}$ cm²/s (cornea)

$D_2 = 7.6 \times 10^{-7}$ cm²/s (contact lens)

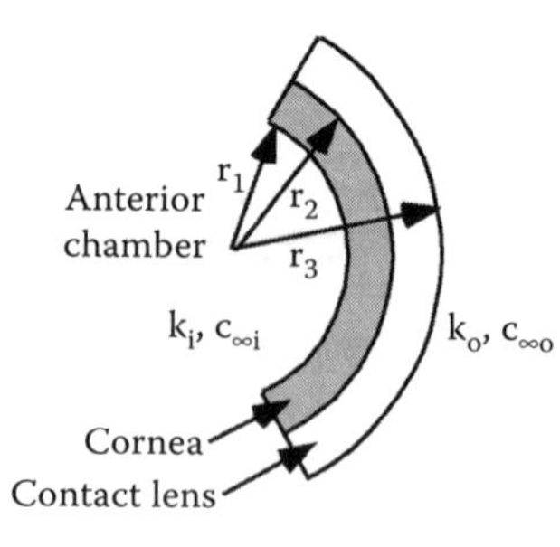

4.5 A layer of soil (pseudo-component B) of thickness 1 m (see figure) that is located above a clay barrier has been contaminated with a toxic gas (A). The concentration of the gas is initially uniform in the soil and equal to 1.0×10^{-5} mol/cm^3. The toxic gas will be removed by making pure air flow over the surface of the soil. The air velocity is such that the convective mass transfer coefficient for the transport of A from the surface of the soil to the bulk air flow is 0.001 cm/s. The diffusivity of the toxic gas in the soil is 0.05 cm^2/s, and its partition coefficient (ratio of concentration of A in air to concentration of A in soil at equilibrium) is 2.5. You might assume that A does not diffuse into the clay barrier.

a. Determine the time necessary to remove 99% of the toxic gas from the soil.
b. Representatives from the company in charge of the clean-up process claim that by increasing the air velocity, the clean-up time can be reduced from the value calculated in part (a) to 2 days. Assess the veracity of the claim.

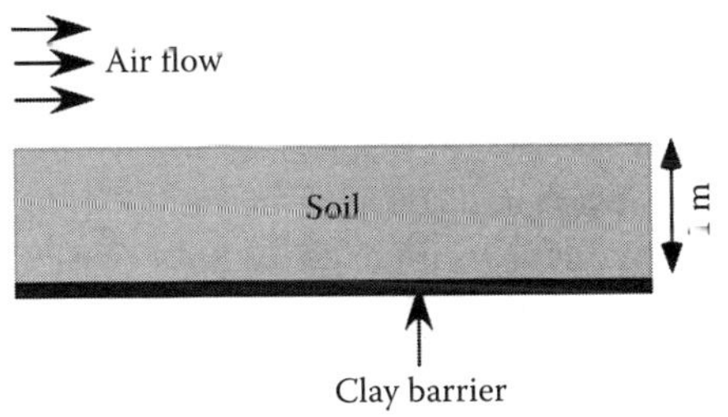

4.6 A process to decaffeinate coffee consists in contacting the coffee beans with super-critical carbon dioxide ($SCCO_2$), free of caffeine, in a well-mixed tank. The coffee beans can be considered to be spheres with a radius of 0.4 cm and a uniform initial concentration of caffeine, and the flow of $SCCO_2$ around the bean is such that convection from the surface to the bulk fluid can be considered very fast. The diffusivity of caffeine (A) in the coffee bean (B) is $D_{AB} = 1.8 \times 10^{-6}$ cm^2/s. The coffee is considered "decaffeinated" when 97% of the original caffeine has been removed. The process has two stages: first, the beans are contacted with $SCCO_2$ for a period of time t_1 at the end of which the caffeine concentration at the center of the bean has fallen to 50% of its original value. The beans are then separated from the $SCCO_2$ and allowed to "rest" for a certain time. During this resting period, the remaining caffeine does not leave the bean and its concentration in the bean eventually becomes uniform. The beans are then contacted with $SCCO_2$ once again for a period of time t_2, until they are "decaffeinated." The two-stage process minimizes the extraction of components from the bean related to flavor and aroma. Determine t_1 and t_2.

4.7 A gas stream exiting a catalytic reaction process contains 1 mol% HCl. Because of the corrosive nature of HCl, it will be removed from the gas by contacting it with a strong basic solution. The gas will be bubbled into the solution. The bubbler used produces uniform spherical bubbles with a 4 mm radius. The acid/base neutralization reaction that occurs when the HCl dissolves in the liquid is so fast that it can be assumed that the liquid immediately in contact with the bubble is free of HCl at all times. The diffusivity of HCl in the gas is 0.12 cm^2/s. How long must each bubble be in contact with the liquid to remove 99.99% of the HCl?

4.8 Polymer spheres (diameter: 1 cm) made up of a mixture of a rubber and an epoxy resin are being considered as controlled release devices to administer sodium nitrate (an essential nutrient) to underground microorganisms that remediate soils contaminated with hydrocarbons. To load the spheres with sodium nitrate, they are contacted with a solution of sodium nitrate in water. After a long time (at equilibrium), the concentration of sodium nitrate in the spheres reaches a uniform value of 3.02×10^{-4} mol/cm^3 while, at the same time, the concentration of sodium nitrate in the liquid is 10×10^{-4} mol/cm^3. After loading, the spheres are placed in the soil where they are completely surrounded by flowing water that contains 10^{-4} mol/cm^3 of sodium nitrate. The convective mass transfer coefficient for transfer of sodium nitrate between the flowing liquid and the spheres is 1×10^{-4} cm/s. The diffusivity of sodium nitrate in the polymer material is 1.3×10^{-5} cm^2/s.

a. Find the time necessary for the spheres to release 80% of the sodium nitrate that they contain when they are initially placed in the soil.
b. Calculate the molar flow rate (moles/s) of sodium nitrate from the sphere to the flowing water at the time when 80% of the sodium nitrate has been released.

4.9 An open, shallow reservoir of a viscous, nonvolatile liquid is located inside a petrochemical plant. We wish to evaluate the unsteady absorption of H_2S by the liquid if this gas is released to the atmosphere. The scenario considered is the sudden increase of the concentration of H_2S in the air over the surface of the reservoir from zero to 0.1 mol%. If this level of contamination is maintained for 30 days,

a. What is the minimum concentration of H_2S in the liquid of the reservoir at the end of this period?
b. What is the maximum concentration?

Additional information:

Mass transfer inside the reservoir occurs by diffusion only. The reservoir can be approximated to a slab 0.1 m deep with a surface 200 m × 200 m. Convective mass transfer in the air flowing over the reservoir can be considered fast.

The diffusivity of H_2S in the liquid at 298 K and 1 atm (reservoir conditions) is 2×10^{-9} m^2/s, and the partition coefficient is $K = 0.4$ (ratio between molar concentration of H_2S in air and molar concentration in the liquid).

4.10 Steel spheres (5 mm radius) that have a uniform concentration of carbon equal to 1.0×10^{-3} mol/cm^3 will be further carburized by exposing them to a gas containing carbon monoxide at 900°C. The concentration of carbon monoxide in the gas is such that the sphere surface is continuously kept at a carbon concentration equal to 4.5×10^{-3} mol/cm^3 in the steel. The diffusivity of carbon in the steel is 2.0×10^{-11} m^2/s.

a. At the end of the process, the spheres will be suddenly cooled in an oil bath, which essentially stops diffusion by binding the carbon to the iron mineral in the steel. If the process is designed to end when the carbon concentration in the steel reaches 3.5×10^{-3} mol/cm^3 at a distance of 2 mm from the sphere surface, determine the time required for the process.
b. What is the total carbon content in a steel sphere at the end of the process (moles)?
c. What is the minimum concentration of carbon in the sphere at the end of the process?

4.11 A solar pond is a shallow layer of water exposed to the sun that is used to generate energy by taking advantage of the solar heating of the water. Efficient operation of the pond necessitates that the water remains stagnant. To avoid natural convection due to solar heating, a salinity concentration gradient is established. This is done by placing at the bottom of the water a layer of pure solid salt (see figure). The following conditions and data apply:

Solubility of salt (sodium chloride) in water: 380 kg/m^3.
Depth of water in the pond: 0.25 m.
Diffusivity of salt in water: 1.2×10^{-9} m^2/s.
Salt does not evaporate into the air.
Initially, the water contains no salt.

a. Find the time (in days) necessary for the concentration of salt to reach 5% of its solubility at the free (top) surface of the water.
b. To find out how much salt should be loaded into the salt layer, determine how much salt (kg) has dissolved into the water at the time calculated in part (a) if the surface area of the pond is 100 m^2.

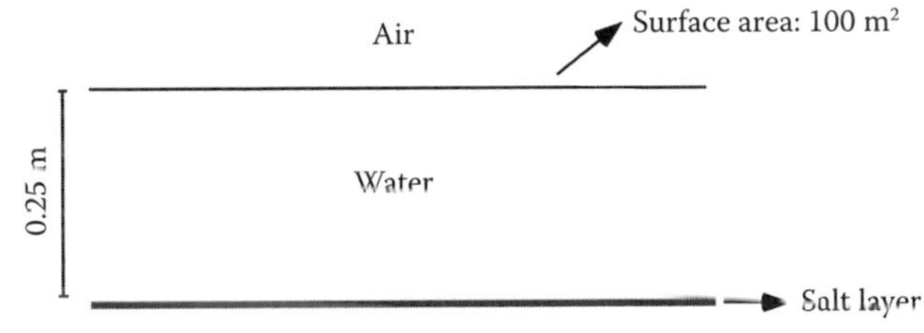

4.12 At the end of a manufacturing process, a flat slab of a polymeric material contains water that must be removed before the material can be used. The initial water concentration in the slab is 3.2×10^{-4} mol/L and the slab has a total thickness of 3 cm. The slab is exposed on both sides to fast-flowing, well-mixed air that contains water with a partial pressure P_w. The air is at

80°C and 1 atm. Under these conditions, water equilibrium between the air and the polymer is given by

$$P_w\,(\text{atm}) = 12.5\ c_{ws}\,(\text{mol/L})$$

where c_{ws} is the concentration of water in the polymeric material. The diffusivity of water in the polymeric material is 1.5×10^{-5} cm²/s.

a. Determine how long it would take to remove 95% of the water from the material if it were exposed to air that is completely dry ($P_w = 0$).
b. It is found that drying the slab so fast produces cracks in the material. For this reason, we wish to increase the drying time calculated in part (a). This will be achieved by increasing the air moisture. Calculate the value that P_w must have to remove 95% of the water from the material in a time that is twice as long as that calculated in part (a).

4.13 Soil (pseudo-component B) has been contaminated with a toxic gas (A). The concentration of the gas is initially uniform in the soil and equal to 1.0×10^{-5} mol/cm³. The toxic gas will be removed by making pure air flow over the surface of the soil. The air velocity is such that convective mass transfer in the air is very fast. The diffusivity of the toxic gas in the soil is 0.05 cm²/s. Treat the soil as a semi-infinite medium. Determine the time necessary to reduce the moles of A to 1% of their initial value from the top 1 m of soil.

5 Diffusive Transport in Reactive Systems

5.1 TREATMENT OF REACTIVE SYSTEMS

Natural and engineered systems often contain chemically reactive constituents. Mass transport of reactive substances differs from that of inert or nonreacting compounds because, by their action, chemical transformations alter or influence the local concentration of species involved with the reaction. Said another way, chemical transformations consume reactants and yield products, so reactive processes are coupled to the transport of individual chemical species. From the perspective of mass conservation, it is instructive to distinguish between two types of chemical reactions:

1. Homogeneous reactions. These are chemical reactions that take place at each point in the continuum. For a species A, they are quantified by the reaction rate R_A (net moles of A produced at a point due to chemical reactions, per unit volume of medium) defined before. In general, R_A depends on temperature and concentrations of all reactants that participate in reactions that produce or consume A.

   A constitutive kinetic expression for R_A must be known. This kinetic expression is used in the continuity equation. For a process that involves diffusion and reaction (no convection), the continuity equation for a dilute species A in a system with constant molar density and diffusivity is equation

$$\frac{\partial c_A}{\partial t} = D_A \nabla^2 c_A + R_A \tag{5.1}$$

   where we are using D_A to denote the diffusivity of A in the mixture. The substitution of the kinetic expression for R_A into this equation will be the first step in the analysis.
2. Heterogeneous reactions. These are reactions that occur only at an interface. In general, the interface will be considered as a boundary of the medium in which mass transfer is analyzed. An example of a heterogeneous reaction is a catalytic reaction occurring on the surface of a solid with the reactants dissolved in a fluid phase in contact with the solid (Figure 5.1). For this reaction to occur, the reactants must be transferred from the fluid to the surface, and the products must be transferred from the surface to the bulk of the fluid.

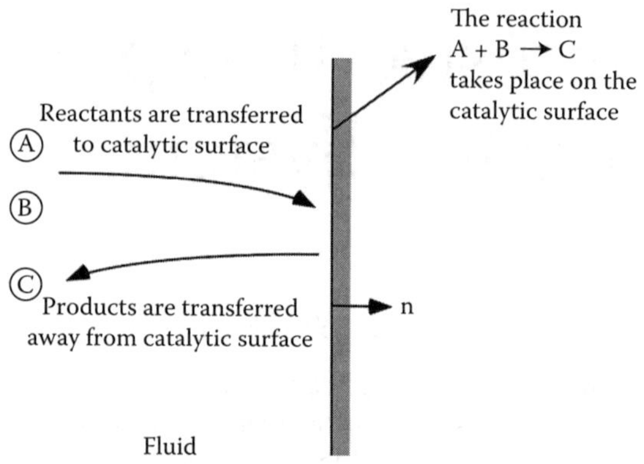

FIGURE 5.1 Example of a heterogeneous reaction on a catalytic surface.

If reactant A is at dilute concentrations, the diffusion equation applies with $R_A = 0$,

$$\frac{\partial c_A}{\partial t} = D_A \nabla^2 c_A, \text{ in the fluid} \tag{5.2}$$

as long as A is not involved in homogeneous reactions. The reaction rate in this case becomes a boundary condition for Equation 5.2. Since the reaction occurs on the solid surface, it is customary to express the reaction rate as a superficial reaction rate, defined as

$$\hat{R}_A = \begin{Bmatrix} \text{Net rate of production of species A on the surface} \\ \text{by chemical reaction per unit area (moles/time area)} \end{Bmatrix} \tag{5.3}$$

Here, we are keeping the convention that the reaction rate is positive when A is produced and negative when A is consumed.

The appropriate boundary condition to be used in the solution of Equation 5.2 comes from a mole balance at the surface: at steady state the moles of A that approach the surface per unit time and per unit surface area must equal the reaction rate of A; that is,

$$\mathbf{N}_A \cdot \mathbf{n} = -\hat{R}_A, \text{ at the surface} \tag{5.4}$$

where **n** is a normal unit vector, pointing away from the fluid. For the example shown in Figure 5.1, if the reaction on the surface is elementary, the heterogeneous reaction rate can be expressed as

$$\hat{R}_A = -\hat{k} c_A c_B \tag{5.5}$$

The reaction on the surface usually occurs between adsorbed species. In that case, it would be better to express this constitutive equation in terms of surface concentrations (i.e., moles of adsorbed species per unit surface area). However, if the surface is at equilibrium with the fluid and partitioning between fluid and surface is linear, Equation 5.5 would be an adequate expression for the heterogeneous reaction rate.

5.2 TRANSPORT AND HOMOGENEOUS REACTION IN A BIOFILM

Some microorganisms present in water attach to available solid surfaces and survive by harvesting nutrients from the water. This happens both in natural systems, such as microorganisms attaching to soil particles, and in engineered systems, where bacteria attached to solid particles are essential in the treatment of wastewater in trickling filters and other bioreactors. As the number of attached microorganisms increases on the surface, they pile up on top of each other and eventually form a solid film on the surface that contains not only the living microorganisms but trapped colloidal particles, cell debris, and secretions from the attached cells. The film can grow to macroscopic sizes and is then called a **biofilm**. Figure 5.2 shows a microphotograph of a biofilm in which the cells have been tagged with fluorescent compounds for their visualization. As the biofilm grows, the only way for nutrients present in the water to reach cells inside the biofilm is by diffusion. The nutrients are consumed in chemical reactions within the cells. This creates a concentration profile of nutrient within the biofilm. If the nutrients are consumed completely before they reach cells close to the solid surface, these cells may die. In this section we will explore the diffusion of nutrients within a biofilm. We will make the assumption that the biofilm is very thin when compared to the radius of curvature of the surface on which it forms, so that we can consider it as a flat slab (Figure 5.3).

To analyze the process of nutrient transport in the biofilm, we first need to know about the kinetics of the nutrient consumption reaction. This reaction is usually modeled by the Michaelis–Menten kinetic expression,

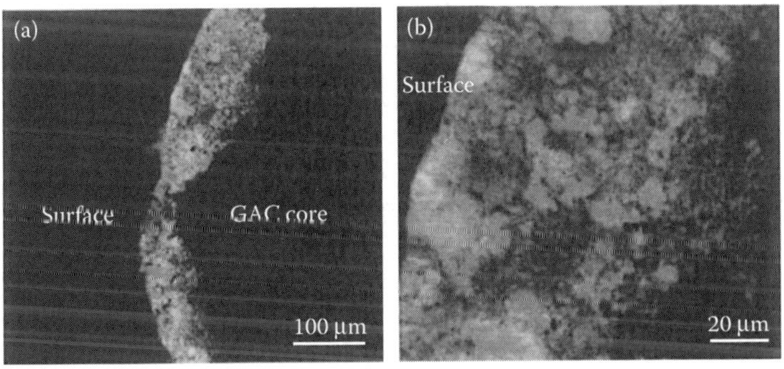

FIGURE 5.2 Biofilm growing on the surface of a granular activated carbon (GAC) particle in a reactor used for the treatment of phenolic wastewater shown at two different magnifications (a and b). Two strains of bacteria (red and green) are active in the biofilm. (From Chen CL et al. Characterization of active microbes in a full-scale anaerobic fluidized bed reactor treating phenolic wastewater, *Microbes Environ*, 24, 144–153, 2009.)

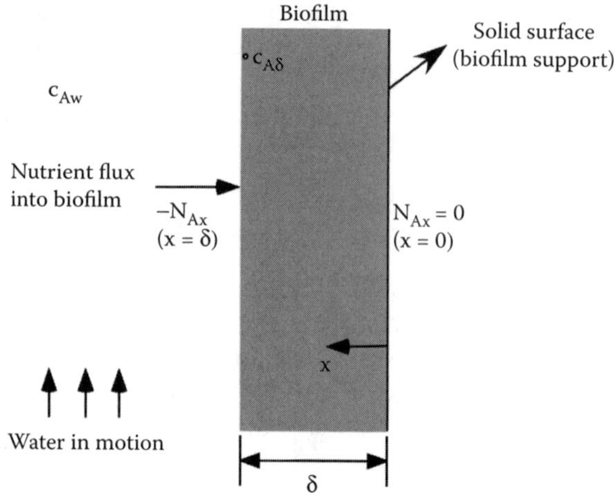

FIGURE 5.3 Scheme of a biofilm. The supporting surface is considered to be inert and, therefore, no flux of nutrient will go into the support solid. The nutrient is present in the bulk water at a concentration c_{AW}. The presence of nutrient in the water will lead to a steady-state biofilm concentration of nutrient $c_{A\delta}$ at the interface.

$$R_A = -\frac{R_{max} c_A}{K_M + c_A} \tag{5.6}$$

This kinetic expression is typical of reactions catalyzed by enzymes. This expression is valid for nutrient consumption for a specific concentration of cells in the biofilm. We will assume that the characteristic time associated with cell concentration changes (and biofilm growth) is much longer than the time required for the nutrient concentration profile to reach steady state.

The kinetic expression has the following interesting limits: first, if c_A is small enough so that $c_A \ll K_M$, the reaction becomes first order in nutrient concentration:

$$R_A = -kc_A, \text{with } k = \frac{R_{max}}{K_M}, \text{valid if } c_A \ll K_M \tag{5.7}$$

On the other hand, for relatively large nutrient concentrations, the reaction becomes zero order:

$$R_A = -R_{max}, \text{valid if } c_A \gg K_M \tag{5.8}$$

Our objectives will be to find the concentration profile of nutrient in the biofilm under steady-state conditions and to find the flux of nutrient from the water into the biofilm. In our solution, we will consider that convective mass transfer in the liquid is fast enough that we can state that the surface of the biofilm is exposed to a concentration c_{Aw} in the liquid. In this case, $c_{A\delta}$ is the concentration of nutrient in the biofilm that is at equilibrium with c_{Aw} and it will be a known value if the partitioning of the nutrient is known.

The problem can be formulated by the one-dimensional, steady-state version of Equation 5.1,

$$D_A \frac{d^2 c_A}{dx^2} + R_A = 0 \tag{5.9}$$

subject to the boundary conditions:

$$\frac{dc_A}{dx} = 0, x = 0 \text{ (no-flux condition)} \tag{5.10}$$

$$c_A = c_{A\delta}, x = \delta \tag{5.11}$$

Consider the case $c_A \ll K_M$, so that the reaction rate is first order (Equation 5.7). The differential Equation 5.9 becomes

$$\frac{d^2 c_A}{dx^2} - \frac{k}{D_A} c_A = 0 \tag{5.12}$$

The solution of this equation is

$$c_A = A \cosh\left(x\sqrt{\frac{k}{D_A}}\right) + B \sinh\left(x\sqrt{\frac{k}{D_A}}\right) \tag{5.13}$$

which leads to

$$\frac{dc_A}{dx} = A\sqrt{\frac{k}{D_A}} \sinh\left(x\sqrt{\frac{k}{D_A}}\right) + B\sqrt{\frac{k}{D_A}} \cosh\left(x\sqrt{\frac{k}{D_A}}\right) \tag{5.14}$$

Applying boundary condition 5.10 leads directly to B = 0. The subsequent use of boundary condition 5.11 in Equation 5.13 yields

$$A = \frac{c_{A\delta}}{\cosh\left(\delta\sqrt{k/D_A}\right)} \tag{5.15}$$

and the solution is

$$c_A = c_{A\delta} \frac{\cosh\left(x\sqrt{k/D_A}\right)}{\cosh\left(\delta\sqrt{k/D_A}\right)} \tag{5.16}$$

We note from the solution that the ratio of the reaction rate constant to diffusivity plays an important role in the concentration profile. To study this effect, we define a dimensionless parameter called the **Thiele modulus** by

$$\phi = \delta\sqrt{\frac{k}{D_A}} \tag{5.17}$$

We also define dimensionless coordinate and concentration by

$$\xi = \frac{x}{\delta} \tag{5.18}$$

$$\Theta = \frac{c_A}{c_{A\delta}} \tag{5.19}$$

so that Equation 5.16 can be expressed as follows:

$$\Theta = \frac{\cosh(\phi\xi)}{\cosh(\phi)} \tag{5.20}$$

The dimensionless concentration profiles are plotted in Figure 5.4 for various values of the Thiele modulus.

From its definition (Equation 5.17), the Thiele modulus squared is a dimensionless ratio between reaction rate and diffusion rate. A high value of ϕ would imply that reaction is fast compared to diffusion into the film. From Figure 5.4 we can see that increasing ϕ leads to a drop in the level of nutrient throughout the film. Notice that for high values of ϕ, the nutrient practically disappears from

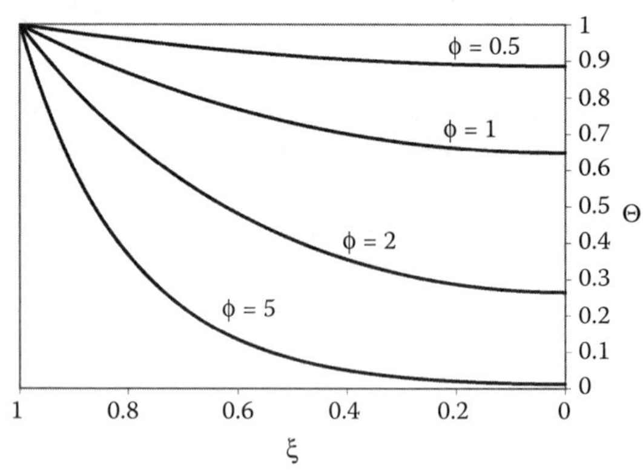

FIGURE 5.4 Dimensionless concentration profiles of a nutrient undergoing first-order consumption in a biofilm, for various values of the Thiele modulus.

sections of the film close to the solid surface. In this case, we say that the film is partially penetrated by the nutrient, which would imply death for cells close to the solid surface.

One of the most important practical aspects of the solution is the quantification of the total amount of nutrient consumed in the film. The moles of nutrient transferred from the liquid into the film per unit time and per unit film surface area are given by the flux of nutrient into the biofilm (Figure 5.3):

$$-N_{Ax}\big|_{x=\delta} = D_A \frac{dc_A}{dx}\bigg|_{x=\delta} = D_A \frac{c_{A\delta}}{\delta}\frac{d\Theta}{d\xi}\bigg|_{\xi=1} \tag{5.21}$$

which, after using Equation 5.20, can be written as

$$-N_{Ax}\big|_{x=\delta} = D_A \frac{c_{A\delta}}{\delta}\phi\tanh(\varphi) \tag{5.22}$$

The effectiveness factor is defined by

$$\eta = \frac{\begin{Bmatrix}\text{Rate of consumption of A (moles} \\ \text{per unit time) in the biofilm}\end{Bmatrix}}{\begin{Bmatrix}\text{Rate of consumption of A that would be obtained} \\ \text{if concentration in the biofilm were uniform at } c_{A\delta}\end{Bmatrix}} \tag{5.23}$$

Let A be the surface area of biofilm; a mole balance in nutrient at steady state states

$$\begin{Bmatrix}\text{Rate of consumption of A (moles} \\ \text{per unit time) in the biofilm}\end{Bmatrix} = -N_{Ax}\big|_{x=\delta} A \tag{5.24}$$

Also, we have

$$\begin{Bmatrix}\text{Rate of consumption of A that would be obtained} \\ \text{if concentration in the biofilm were uniform at } c_{A\delta}\end{Bmatrix} = kc_{A\delta}A\delta \tag{5.25}$$

(since $A\delta$ is the total volume of biofilm).

Equation 5.23 can be written as

$$\eta = \frac{-N_{Ax}\big|_{x=\delta}}{kc_{A\delta}\delta} \tag{5.26}$$

Substituting Equation 5.22 into Equation 5.26 and using the definition of the Thiele modulus yields

$$\eta = \frac{\tanh(\phi)}{\phi} \tag{5.27}$$

This equation is represented in Figure 5.5. Two interesting extreme limits of behavior can be observed in Figure 5.5. First, as $\phi \to 0$ we see that $\eta \to 1$ (it can be shown that $\tanh \phi \to \phi$ as $\phi \to 0$). As the Thiele modulus decreases, the concentration profile in the biofilm becomes flatter (Figure 5.4), which means that the consumption reaction occurs at concentrations closer to the surface concentration throughout the film. In this case, the rate of consumption of nutrient per unit surface area of biofilm is given by (Equation 5.26 with $\eta = 1$)

$$-N_{Ax}\Big|_{x=\delta} \approx kc_{A\delta}\delta, \quad \text{for } \delta < 0.2 \tag{5.28}$$

On the other hand, at high values of the Thiele modulus ($\phi \to \infty$), it is possible to show that $\tanh \phi \to 1$ and thus $\eta \to 1/\phi$. We can see from Figure 5.4 that this behavior is obtained for $\phi \geq 2$. In terms of Equation 5.22, this limit means that

$$-N_{Ax}\Big|_{x=\delta} \approx D_A \frac{c_{A\delta}}{\delta}\phi, \quad \text{for } \phi \geq 2 \tag{5.29}$$

Substituting Equation 5.17 into this equation, we find

$$-N_{Ax}\Big|_{x=\delta} \approx c_{A\delta}\sqrt{kD_A}, \quad \text{for } \phi \geq 2 \tag{5.30}$$

High values of the Thiele modulus imply that the reaction rate is fast compared to the diffusion rate. Under these circumstances, one would expect diffusion to

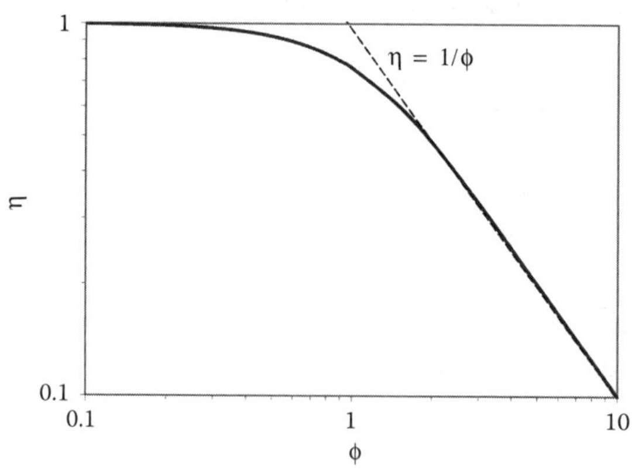

FIGURE 5.5 Effectiveness factor for first-order consumption of nutrient in a biofilm.

dominate, since it will be the slow step. However, diffusion in this case is motivated by the existence of a chemical reaction (if $k = 0$, there would be no concentration gradients inside the film), so that the role of diffusion becomes to provide the reactant to the biofilm, and both k and D_A play a role in determining the consumption rate.

5.3 DIFFUSION WITH HETEROGENEOUS CHEMICAL REACTION

Here, we consider a chemical reaction that occurs on the surface of a solid that is in contact with a fluid phase. The most common example is the case in which the reactants are in the fluid phase and the solid surface contains a catalyst that is insoluble in the fluid, so that the reactants must adsorb onto the surface before the reaction can take place. Another common example of this situation occurs when at least one of the reactants is part of the solid phase and it is insoluble in the fluid. In this case, the reaction will proceed when the molecules of the reactants in the fluid phase approach the solid surface and contact the solid reactant. An example of such a reaction is the oxidation of solid carbon to carbon monoxide in the presence of oxygen:

$$2C_{(s)} + O_{2(g)} \rightarrow 2CO_{(g)} \tag{5.31}$$

Let $A = O_2$ and $B = CO$. If this reaction is elementary and irreversible, and the solid is pure carbon, we would expect a kinetic rate given by

$$\hat{R}_A = -\hat{k}c_{As} \tag{5.32}$$

where c_{As} is the concentration of oxygen in the fluid that is in contact with the solid surface where carbon is pure. This kinetic rate expression would apply to any kind of first-order consumption reaction at the surface. An additional example could be a reaction like

$$A_{(g)} \rightarrow 2B_{(g)} \tag{5.33}$$

that occurs on the solid surface due to the presence of a catalyst. The rate expression for this equation could well be Equation 5.32, and the stoichiometry of the gas-phase reactants would be the same as those of reaction 5.31. Let us consider reaction 5.33 as a model for analysis. The case in question will be the occurrence of this reaction on a solid surface exposed to a moving gas. We will assume that the gas is made up from A and B only (hence we cannot assume that any component is present at dilute conditions), and that there is convective mass transfer in the gas phase. The most critical mass transfer process will occur in the vicinity of the surface, where we will hypothesize that there exists a thin film (thickness δ) in which transport occurs only by diffusion from the bulk of the fluid to the surface (Figure 5.6). The mole fraction of A in the bulk of the fluid, $x_{A\delta}$, is known, and the mole fraction of B will be $x_{B\delta} = 1 - x_{A\delta}$. The consumption of A on the solid surface will lower the concentration of A with

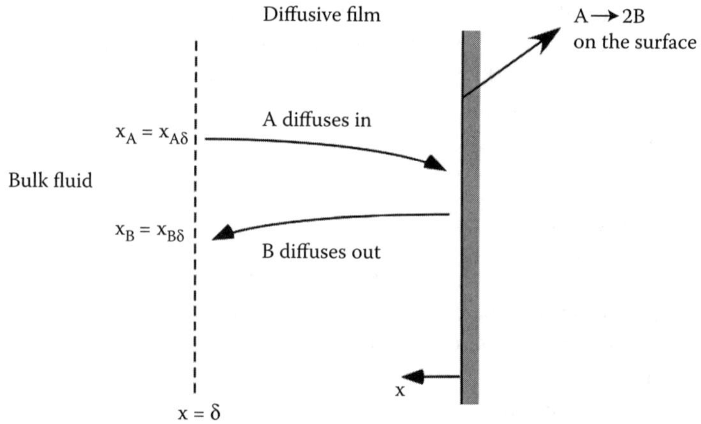

FIGURE 5.6 A diffusion film in the vicinity of a solid surface on which a chemical reaction occurs.

respect to the bulk fluid, thereby creating a concentration gradient that will induce diffusion of A from the bulk toward the surface. Similarly, the creation of B on the surface will raise its concentration in the vicinity of the solid, creating a concentration gradient that will induce diffusion of B from the surface to the bulk fluid. In the analysis that follows, we will assume that the process has reached steady state.

Within the diffusion film there will be no homogeneous chemical reaction ($R_A = 0$). If there is only transport in the x direction, the continuity equation for component A simplifies to

$$\frac{dN_{Ax}}{dx} = 0 \tag{5.34}$$

Similarly, for component B we have

$$\frac{dN_{Bx}}{dx} = 0 \tag{5.35}$$

Equations 5.34 and 5.35 imply that the fluxes of both components are constant in the diffusion film. The two fluxes are related by the stoichiometry of the reaction: at steady state, the stoichiometry of reaction 5.33 implies that for each mole of A that approaches the surface and is, therefore, consumed in the reaction, two moles of B must be produced. This leads to the following **stoichiometric** balance on the surface:

$$N_{Bx}\Big|_{x=0} = -2N_{Ax}\Big|_{x=0} \tag{5.36}$$

Note that, in this case, specifying $x = 0$ is unnecessary since the fluxes are the same everywhere. The minus sign in Equation 5.36 reflects the fact that the fluxes must have opposite directions.

Now, we write the flux equation for component A in the x direction:

$$N_{Ax} = x_A(N_{Ax} + N_{Bx}) - cD_{AB}\frac{dx_A}{dx} \tag{5.37}$$

Making use of Equation 5.36 and rearranging yields

$$N_{Ax}(1 + x_A) = -cD_{AB}\frac{dx_A}{dx} \tag{5.38}$$

Separating and integrating leads to

$$-\frac{N_{Ax}}{cD_{AB}}\int_{x}^{\delta} dx = \int_{x_A}^{x_{A\delta}} \frac{dx_A}{(1 + x_A)} \tag{5.39}$$

which implies

$$-\frac{N_{Ax}}{cD_{AB}}(\delta - x) = \ln\left(\frac{1 + x_{A\delta}}{1 + x_A}\right) \tag{5.40}$$

This equation yields the mole fraction profile, $x_A(x)$, once N_{Ax} is known. To find the flux, we make use of the reaction rate. A mole balance of A on the surface states that the moles of A per unit time and surface area that reach the surface must equal the moles of A per unit time and surface area consumed in the reaction. Using the rate expression 5.32 and letting $c_{As} = cx_{As}$, the balance is

$$-N_{Ax} = \hat{k}cx_{As} \tag{5.41}$$

Substituting Equation 5.41 into Equation 5.40 leads to

$$\frac{\hat{k}}{D_{AB}}x_{As}(\delta - x) = \ln\left(\frac{1 + x_{A\delta}}{1 + x_A}\right) \tag{5.42}$$

Evaluating this equation at $x = 0$ results in

$$\frac{\hat{k}\delta}{D_{AB}}x_{As} = \ln\left(\frac{1 + x_{A\delta}}{1 + x_{As}}\right) \tag{5.43}$$

From this equation, x_{As} can be calculated (notice that a numerical solution would be required since Equation 5.43 cannot be solved explicitly for x_{As}). Once x_{As} is known, the mole fraction profile can be obtained from Equation 5.42:

$$x_A = (1 + x_{A\delta})\exp[-\alpha x_{As}(\delta - x)] - 1 \tag{5.44}$$

where

$$\alpha = \frac{\hat{k}\delta}{D_{AB}} \tag{5.45}$$

This dimensionless parameter is similar to the Thiele modulus used in the previous section in the sense that it represents a ratio between reaction and diffusion rates. It is interesting to explore the effects of extreme values for α on the solution found. First, consider $\alpha \ll 1$ (mathematically we explore the limit as $\alpha \to 0$). From Equation 5.44, we can see that this limit yields the profile

$$x_A \equiv x_{A\delta}, \alpha \ll 1 \tag{5.46}$$

that is, the mole fraction profile is flat. The consumption rate of A can then be found from Equation 5.41 to be

$$-N_{Ax} = \hat{k}cx_{A\delta}, \alpha \ll 1 \tag{5.47}$$

which indicates that the rate of consumption of A is independent of the diffusivity. Physically, this limit implies that the reaction rate is so slow compared to diffusion that the film is perfectly mixed by the diffusion process and hence diffusion ceases to play a role. In other words, the reaction on the surface is the slow step that controls the process and we say that the process is **reaction-controlled**.

On the other hand, when $\alpha \gg 1$ (mathematically, we explore the limit as $\alpha \to \infty$), Equation 5.43 implies that $x_{As} \to 0$ (the reaction is so fast that all the A that gets to the surface is consumed). In this case, the consumption rate can be obtained by letting $x = 0$ in Equation 5.42, which yields

$$N_{Ax} = -\frac{cD_{AB}}{\delta}\ln(1 + x_{A\delta}) \tag{5.48}$$

In this case, the consumption flux is independent of the reaction rate and we say that the process is **diffusion-controlled**.

5.4 ENHANCED (FACILITATED) DIFFUSION

Enhanced or facilitated diffusion refers to a particular case of transport that results in mass transfer rates for a solute that are appreciably higher than what would be expected. Facilitated transport is not limited to diffusion, and usually (although not always) is accompanied by chemical reactions. Examples of facilitated transport are

1. Transport of oxygen from lungs to tissue. Owing to the relatively low solubility of oxygen in blood (or any aqueous liquid), its transport from the lungs to tissue around the body would not be adequate if it were to occur by direct dissolution. What occurs is a complexation reaction in which oxygen forms a complex with hemoglobin (oxyhemoglobin). The complex is transported by convection (blood flow in arteries) to the tissue, where molecular oxygen is released from the complex.
2. Groundwater transport of low-solubility contaminants by colloidal matter. Natural colloids (particles in the sub-micron or nanoscale) are ubiquitous in groundwater. Insoluble contaminants adsorbed or deposited in the soil might adsorb on the colloidal particles and thus get transported with the main groundwater flow. Examples of this are insoluble organics being transported by humic colloids and radioactive contaminants transported by colloidal soil particles.
3. Transport through the cellular membrane. The selectivity of cell membranes to let chemical species be transported into or out of the cell is related to its capacity for facilitated transport of specific solutes. This control is exerted by binding agents that produce complexes with certain solutes.
4. Drug delivery systems. The concept of a drug delivery system and controlled-release systems is closely related to facilitated transport. The idea is to design a medium that can transport the drug to the specific tissue where it is required, having controllable release rates.

Some of the examples above rely on chemical reactions (1, 3, and some applications of 4), while others rely on other mechanisms, such as adsorption (2) or absorption (some applications of 4).

In this section, we consider an example of enhanced diffusion through a flat membrane. The situation to consider is the transport of a species (A) between two fluids separated by a solid membrane, in which A is completely insoluble. To accomplish transport of A, the membrane contains a species C that reacts with A to form a complex B that is soluble in the membrane. The reaction is reversible and occurs at the membrane–fluid interface (Figure 5.7),

$$A + C \underset{\hat{k}_2}{\overset{\hat{k}_1}{\rightleftarrows}} B \tag{5.49}$$

We will assume that this reaction is elementary, and that there is always an excess of C present in the membrane. The rate of heterogeneous reaction can be expressed as

$$\hat{R}_A = -\hat{R}_B = -\hat{k}_1 c_A + \hat{k}_2 c_B, \text{ at membrane/fluid interface} \tag{5.50}$$

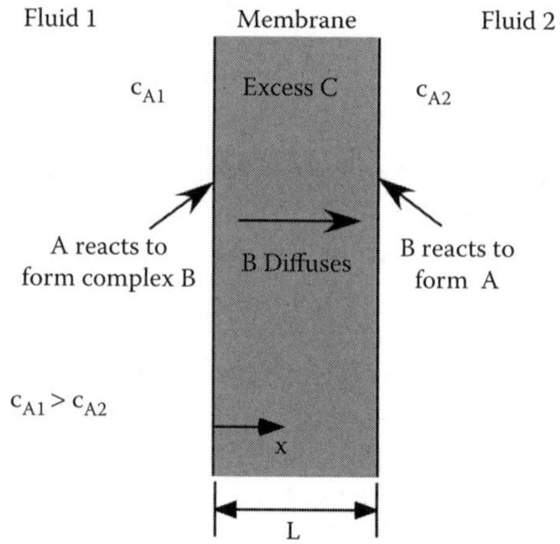

FIGURE 5.7 Facilitated diffusion through a membrane: A is insoluble in the membrane but reacts with C (x = 0) to produce a soluble complex B that diffuses through the membrane and releases A on the other side (x = L). The net effect is the transport of A from one fluid phase (1) to the other (2).

where c_A is the concentration of A in the fluid and c_B is the concentration of complex in the membrane. We will consider that the process is at steady state, and that B is insoluble in the fluids surrounding the membrane and is present in dilute conditions in the membrane ($c_B \ll c$).

Since the chemical reaction occurs only at the interfaces (x = 0, L), species B is only subject to diffusive transport in the membrane. The steady-state continuity equation for B therefore reduces to (assuming that diffusion occurs only in the x direction)

$$\frac{dN_{Bx}}{dx} = 0 \Rightarrow N_{Bx} \text{ is uniform} \tag{5.51}$$

Let c_{B0} and c_{BL} be the concentrations of B at x = 0 and L, respectively (these values are unknown). The simple diffusion of B through the membrane leads to a linear concentration profile with a diffusive flux given by

$$N_{Bx} = -D_B \frac{c_{BL} - c_{B0}}{L} \tag{5.52}$$

where D_B is the diffusivity of B in the membrane. On the other hand, the flux of B is controlled by the occurrence of the heterogeneous chemical reaction 5.49 at the planes x = 0 and L. Mole balances of B on those planes lead to

$$N_{Bx} = \hat{R}_B, \ x = 0 \tag{5.53}$$

(moles of B produced equal moles of B transported by diffusion), and

$$N_{Bx} = -\hat{R}_B, \; x = L \tag{5.54}$$

(moles of B transported equal moles of B consumed). Using Equations 5.50 and 5.52 in Equations 5.53 and 5.54 leads to

$$D_B \frac{c_{B0} - c_{BL}}{L} = \hat{k}_1 c_{A1} - \hat{k}_2 c_{B0} \tag{5.55}$$

and

$$D_B \frac{c_{B0} - c_{BL}}{L} = -\hat{k}_1 c_{A2} + \hat{k}_2 c_{BL} \tag{5.56}$$

Rearranging these two equations leads to

$$\left(\frac{D_B}{L} + \hat{k}_2\right) c_{B0} - \frac{D_B}{L} c_{BL} = \hat{k}_1 c_{A1} \tag{5.57}$$

$$\frac{D_B}{L} c_{B0} - \left(\frac{D_B}{L} + \hat{k}_2\right) c_{BL} = -\hat{k}_1 c_{A2} \tag{5.58}$$

This is a system of two equations with two unknowns (c_{B0} and c_{BL}). The solution for c_{B0} is

$$c_{B0} = \frac{\hat{k}_1 c_{A1}\left((D_B/L) + \hat{k}_2\right) + \hat{k}_1 c_{A2}(D_B/L)}{\left((D_B/L) + \hat{k}_2\right)^2 - (D_B/L)^2} \tag{5.59}$$

Ultimately, our goal in this problem is to calculate the flux of A that is transferred from fluid 1 to fluid 2. Let us call this flux N_{At}. This flux can be calculated by a stoichiometric balance at either of the two interfaces. For example, at x = 0, we can state that the amount of A that goes into the membrane is equal to the amount of complex (B) generated, since the stoichiometry is 1:1; that is,

$$N_{At} = N_{Bx} = \hat{k}_1 c_{A1} - \hat{k}_2 c_{B0} \tag{5.60}$$

Substituting Equation 5.59 into Equation 5.60 leads to (after manipulations)

$$N_{At} = \frac{\hat{k}_1}{\hat{k}_2 + (2D_B/L)} D_B \frac{c_{A1} - c_{A2}}{L} \tag{5.61}$$

This represents the rate of facilitated diffusion of A through the membrane. Notice that, as expected, this rate depends on the reaction kinetics, the diffusivity of B through the membrane, and the membrane thickness.

Equation 5.61 can be analyzed for relevant extreme cases as follows. First of all, consider the definition of equilibrium constant for reaction 5.49:

$$K_{eq} = \frac{\hat{k}_1}{\hat{k}_2} \tag{5.62}$$

Equation 5.61 can be rewritten as

$$N_{At} = \frac{K_{eq}}{1 + (2D_B / \hat{k}_2 L)} D_B \frac{c_{A1} - c_{A2}}{L} \tag{5.63}$$

Now, consider the following extreme cases:

1. Fast reaction, slow diffusion, $(2D_B / \hat{k}_2 L) \gg 1$. Equation 5.63 reduces to

$$N_{At} = K_{eq} D_B \frac{c_{A1} - c_{A2}}{L} \tag{5.64}$$

   The flux resembles a diffusive flux for A, but the diffusivity is that of the complex (D_B). Note that the enhanced transport of A is accelerated, the larger the equilibrium constant becomes.

2. Slow reaction, fast diffusion, $(2D_B / \hat{k}_2 L) \gg 1$. Equation 5.63 reduces to

$$N_{At} = \frac{\hat{k}_1}{2}(c_{A1} - c_{A2}) \tag{5.65}$$

   In this case, diffusion is so fast that it becomes irrelevant (Equation 5.65 is also obtained by letting $c_{B0} = c_{BL}$ in the reaction rate analysis).

PROBLEMS

5.1 A solid spherical particle (pure A) of radius R dissolves in a large, stagnant pool of liquid (B). The diffusivity of dissolved A in the liquid is D_{AB}. To enhance the dissolution process, the liquid contains a chelating agent that will react with dissolved A to remove it from solution. The reaction follows first-order kinetics $R_A = -kc_A$. The dissolution process can be considered quasi steady, so that in the liquid $c_A = c_A(r)$. The solubility of A in the liquid is c_{As}.

a. Write the differential equation and boundary conditions whose solution yields the concentration profile $c_A(r)$.
b. Use the change of variables $c_A = u/r$ to show that the differential equation becomes

$$\frac{d^2u}{dr^2} - \frac{k}{D_{AB}}u = 0$$

c. Solve the problem and find $c_A(r)$.
d. From the solution found, determine the molar flux of A at the sphere surface (dissolution flux).
e. To quantify the enhancement in dissolution flux provided by the presence of the chelating agent, find the ratio between the dissolution flux found in (d) and the dissolution flux in the absence of a chelating agent.

5.2 One of the limiting steps in the transport of oxygen from blood to tissues is the diffusion of oxygen through the tissue. Consider the idealized model shown in the figure in which a capillary transports blood to a cylindrical tissue. The concentration of oxygen (A) in the blood ($c_{Ab} = 4.05 \times 10^{-8}$ mol/cm^3) can be assumed to be uniform in the capillary both radially and axially. In the tissue ($R_c < r < R$) the oxygen diffuses radially at steady state and is consumed by aerobic metabolism in the tissue cells at a uniform rate $R_A = -1 \times 10^{-8}$ mol/cm^{3}s. The external surface of the tissue ($r = R$) is impermeable to oxygen. The radii of the capillary and tissue cylinders are: $R_c = 1.5 \times 10^{-4}$ cm and $R = 50 \times 10^{-4}$ cm. The diffusivity of oxygen in the tissue is 2.0×10^{-5} cm^2/s. The partition coefficient of oxygen between blood and tissue is 1.

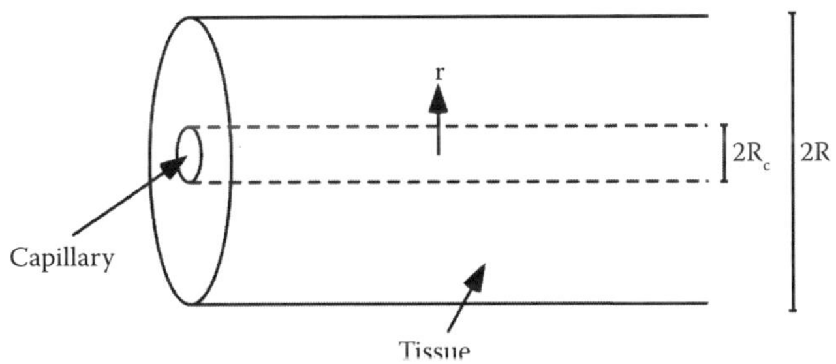

a. Determine the concentration of oxygen at the external surface of the tissue.
b. Calculate the molar flux of oxygen that flows from the capillary to the tissue.

5.3 An airborne spherical microorganism consumes oxygen by respiration. The consumption of oxygen occurs uniformly through the microorganism, and it can be assumed to follow a zeroth-order kinetic rate expression: $R_A = -k$ ($A = O_2$). Laboratory studies have determined that a specific microorganism with radius $R = 0.01$ cm consumes 1×10^{-8} mol of O_2/h in atmospheric air at 20°C under steady-state conditions. The diffusivity of oxygen in the microorganism is $D_{AB} = 1 \times 10^{-5}$ cm^2/s. The solubility of oxygen in the microorganism is $c_{As} = 1.4 \times 10^{-6}$ mol/cm^3, when exposed to atmospheric air. If mass transfer from the surrounding air to the surface of the microorganism is very fast, determine:

a. The kinetic rate constant, k.
b. The concentration of oxygen at the center of the microorganism, c_{Ac}.

5.4. The accelerated increase of carbon dioxide levels in the atmosphere is believed to be the consequence of anthropogenic CO_2 emissions and it has been linked to global warming. Another detrimental consequence of higher CO_2 concentrations in the lower atmosphere is the acidification of the oceans. If the pH of the oceans becomes low enough, a large number of species that rely on deposition of calcium carbonate for their survival (such as corals) will be endangered. The absorption of CO_2 into the ocean is enhanced by its homogeneous reaction with hydroxyl ions in the liquid to form soluble carbonate species:

$$CO_2 + OH^- \rightarrow HCO_3^-, R_A = -kc_Ac_B$$

where $A = CO_2$ and $B = OH^-$, and k is known. As a first step in modeling ocean uptake of CO_2, consider the absorption of CO_2 into a static deep pool of liquid. The liquid will be a relatively high-pH aqueous solution, so that the concentration of OH^- (c_B) is known and, being in excess, uniform. The liquid is exposed to an atmosphere with a known partial pressure of CO_2 (P_A). The diffusivity of CO_2 in the liquid is D_A, and equilibrium between gas and liquid is given by Henry's law, $P_A = Hc_A$, where H is known. Deep into the liquid, it can be assumed that $c_A \approx 0$. At any point in the liquid, CO_2 will be dilute.

a. Find the steady-state concentration profile of CO_2 in the liquid, $c_A(z)$.
b. Derive an expression for the flux of CO_2 that enters the liquid at the surface.

5.5 Radon (specifically, the isotope ^{222}Rn) is a radioactive element that occurs naturally in certain types of soil. It is produced from the radioactive decay of uranium, so that it is present at relatively high levels in soils at uranium mining sites. Even low concentrations of radon in the air have shown to cause lung cancer. To mitigate mine workers exposure to radon in an area that has relatively high radon release, someone has proposed to flood the area with a pool of water of depth L (see figure). Because of the relatively slow diffusion of radon through the water and its relatively fast decomposition due to radioactive decay, it is expected that the pool of water would substantially decrease the rate of radon release to the air.

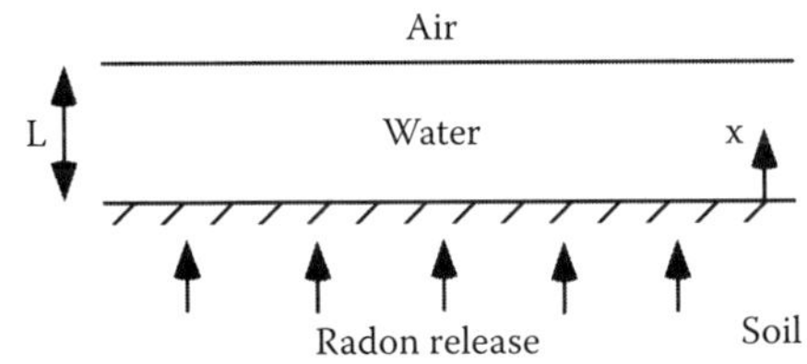

Let N_{Ax0} and N_{AxL} be the molar fluxes of radon from the soil to the water and from the water to the air, respectively. Find the depth of the pool of water (L) that would result in a 99% flux attenuation; that is: $N_{AxL} = 0.01N_{Ax0}$. The diffusivity of radon in water is $D_{Aw} = 1.2 \times 10^{-9}$ m²/s. The rate of radioactive decomposition of radon in water is first order in radon concentration with a rate constant $k = 2.1 \times 10^{-6}$ s⁻¹. You may assume that the radon concentration in the air over the water surface is negligibly small and that the process is at steady state.

5.6 The elementary isomerization reaction $A \rightarrow C$ will be carried out in a soft gel that is attached to a flat solid surface. The reactant A is available as part of a gas that also contains oxygen. The reaction only can proceed in the presence of a catalyst that is soluble in the gel but that will oxidize and deactivate if exposed to oxygen. To avoid catalyst oxidation, we propose to impregnate with a catalyst only half of the gel layer ($0 < x < \delta$, see figure) while maintaining a catalyst-free top layer ($\delta < x < 2\delta$) that will isolate the catalyst from the gas. In this arrangement, A will diffuse from the gas into the gel and will react to produce C in the reactive layer. The reaction product C will then diffuse toward the gas, since the solid surface (x = 0) is impermeable to both A and C. Both A and C are present at dilute concentrations throughout the gel. The partition of A between the gas and the gel is given by Henry's law: $P_A = Hc_A$, with H known. The diffusivity of A in the gel is D_A. Diffusion of the catalyst into the nonreactive layer can be neglected.

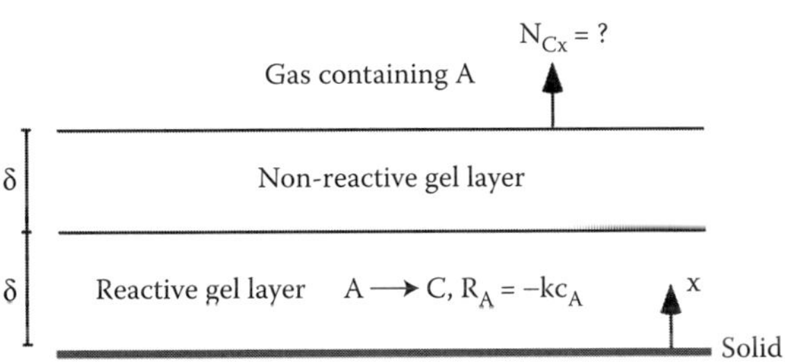

a. Find the steady-state concentration profile of A in the reactive gel layer, in terms of H, D_A, k, δ, and the partial pressure of A in the gas, P_A.
b. Find an expression to calculate the molar flux of C that will be transferred to the gas at steady state (i.e., N_{Cx} at $x = 2\delta$).

5.7 A bioreactor is loaded with spherical support particles of radius R, each of which contains a uniform distribution of aerobic cells. The oxygen concentration on the surface of the particles is c_{A0}. The steady-state oxygen concentration profile in the particles is needed in the design of this reactor to ensure that cells near the center have enough oxygen to live. Experimental measurements show that the respiration rate of the cells is first order in the oxygen concentration, $R_A = -kc_A$. The diffusivity of oxygen in the particle is D_A.

a. Find the steady-state oxygen concentration profile in the spherical particle, $c_A(r)$.

b. Find an expression to calculate the moles of oxygen consumed per unit time in the particle, W_A.

c. Determine a constraint that must be satisfied to ensure that $c_A\big|_{r=0} > 0.1\, c_{A0}$.

5.8 Compound semiconductors are materials used in solar cells, lasers, and transistors. Such devices are produced by growing very thin films of controlled purity over large areas (the films are sometimes a few molecules thick). One method to deposit a film with these characteristics is chemical vapor deposition, in which a carrier gas, such as H_2, conveys reactive vapors, such as arsine (AsH_3), and an organometallic species, such as trimethylgallium ($(CH_3)_3Ga$, TMG) to a heated surface. Thermal decomposition occurs at the surface, leading to the deposition of a film of gallium arsenide (GaAs):

$$AsH_{3(g)} + (CH_3)_3Ga_{(g)} \rightarrow GaAs_{(s)} + 3CH_{4(g)}$$

This reaction is so fast and arsine is usually in such an excess that the process is controlled by the diffusion of TMG between the bulk of the gas phase and the flat surface.

A GaAs film is deposited on a square flat plate with sides 20 cm long. The pressure and temperature of the process are 1 atm and 600 K, respectively. At these conditions, the diffusivity of TMG in the gas (which is mostly H_2) is 1 cm^2/s. TMG is present in the gas at a partial pressure of 7×10^{-5} atm at a distance 0.1 cm from the surface. You may assume that diffusion in the gas is pseudo steady (i.e., neglect the increasing thickness of the forming film of GaAs).

a. Determine the rate at which the film is formed in mol GaAs deposited/s.

b. Determine the molar flux of CH_4 that diffuses away from the surface. The diffusivity of CH_4 in the gas is 2 cm^2/s.

5.9 A bimolecular reaction

$$A + B \rightarrow AB$$

is carried out in a solid membrane by setting the concentration of A at one end of the membrane ($x = 0$) to c_0 and the concentration of B at the other end ($x = \delta$) to the same value c_0. Both A and B are dilute in the membrane and their diffusivities are known (D_A and D_B). The reaction is so fast that A and B are completely consumed when they approach each other, as shown in the figure. Find an expression to calculate the location of the reaction front, δ_R.

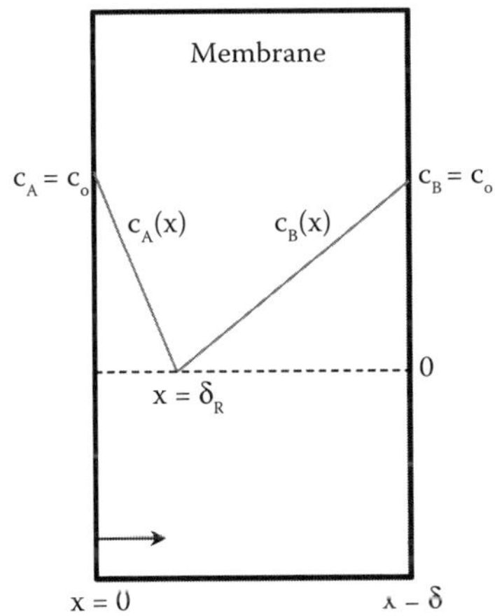

5.10 A soil contaminated with a volatile compound (A) releases it into the atmosphere at a constant known molar flux N_{A0}. In the quiescent atmosphere, the contaminant is destroyed by solar photolysis following a first-order process, $R_A = -kc_A$. The diffusivity of the compound in air is D_A and it is known that it is present in the air at relatively low concentrations and only because of its release from the soil. Find the steady-state concentration profile in the atmosphere, $c_A(z)$.
If the compound becomes toxic at concentrations $c_A > c_{Am}$, find an expression to calculate the height ($z = H$) above which it would be safe to breathe the air. If the release flux is increased 10-fold: $N_{A0,2} = 10\, N_{A0,1}$, calculate the difference in the toxic limit heights, $H_2 - H_1$.

5.11 A solid film made of chemical species S is formed by the reaction between a solid (A) and a compound (B) dissolved in a liquid phase (see figure). As the film grows, A dissolves in it and diffuses to the solid/liquid interface ($x = \delta$) where it reacts with B. Species B is insoluble in the solid formed by the produced compound S, and both A and S are insoluble in the liquid that carries B. The liquid is well mixed so that the concentration of B at the solid/liquid interface (c_B) is known and constant. The reaction rate is given by

$$\hat{R}_A = -\hat{k} c_B c_{A\delta}$$

where $\hat{R}_A$ is in moles of A per unit time and surface area of interface, and $c_{A\delta}$ is the concentration of A in the solid at the interface.

Other known parameters are: the solubility of A in the solid film (c_{Ae}), the diffusivity of A in S (D_{As}), the molar density of the solid (c_s)—consider that $c_{Ae} \ll c_s$. Even though the film grows continuously, the diffusion process of A in the film can be considered to be at steady state at all times. Find expressions to calculate:

a. The interfacial concentration of A, $c_{A\delta}$, in terms of the information given.
b. The molar flux of A that dissolves from the solid (pure A) surface.
c. The film thickness as a function of time, $\delta(t)$, if $\delta(0) = 0$.

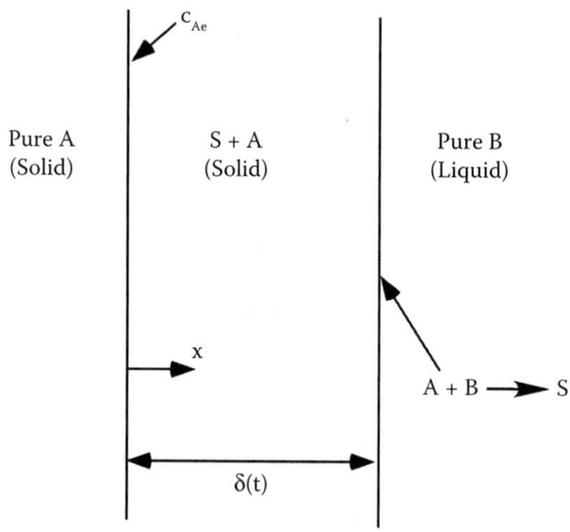

6 Convective Transport

6.1 THE ROLE OF MOMENTUM BALANCES IN CONVECTIVE TRANSPORT

The convective–diffusion equation derived in Chapter 3 describes mass transport in systems when both convection and diffusion have important roles in the transport of a chemical species A,

$$\frac{\partial c_A}{\partial t} + \mathbf{v} \cdot \nabla c_A = D_A \nabla^2 c_A + R_A \tag{6.1}$$

It is important to recall that this equation is valid for binary systems or multicomponent systems in which diffusion of A can be modeled by Fick's law (e.g., systems in which A is dilute), with uniform diffusivity. The use of Equation 6.1 in the formulation of a mass transfer problem requires knowledge of the velocity vector, **v**. Prediction of the velocity field necessitates the use of momentum conservation to derive the point equations that describe fluid motion. In most practical applications, the momentum balance can be solved to find the velocity field independently of the mass transfer process. That is, the fluid mechanics and mass transfer processes are decoupled.

In the next section, we will provide an introduction to the use of momentum balances to determine velocity fields. A more extensive treatment of momentum balances can be found in Bird et al. (2007).

6.2 MASS AND MOMENTUM BALANCES

The study of momentum transport is based on the application of mass and linear momentum conservation. In Chapter 2, we showed how to perform total mass balances on arbitrary control volumes. Here, we will treat linear momentum balances. The general principle of linear momentum conservation applied to an arbitrary control volume, $V_a(t)$ (Figure 2.1), states

$$\begin{Bmatrix}\text{The time rate of change of}\\ \text{the total linear momentum}\\ \text{in an arbitrary volume}\end{Bmatrix} = \begin{Bmatrix}\text{Rate at which linear momentum}\\ \text{enters the arbitrary volume}\end{Bmatrix} - \begin{Bmatrix}\text{Rate at which linear momentum}\\ \text{exits the arbitrary volume}\end{Bmatrix} + \begin{Bmatrix}\text{Total force exerted by the surroundings}\\ \text{on the arbitrary volume}\end{Bmatrix} \tag{6.2}$$

The linear momentum of a fluid particle is the product between the mass and the particle velocity. Thus, for an element of fluid dV, the linear momentum is $\rho\mathbf{v}\,dV$, and the total linear momentum in the arbitrary volume is found by integrating this quantity over $V_a(t)$, so that the left-hand side of Equation 6.2 can be expressed as

$$\begin{Bmatrix}\text{The time rate of change of}\\ \text{the total linear momentum}\\ \text{in an arbitrary volume}\end{Bmatrix} = \frac{d}{dt}\int_{V_a(t)} \rho\mathbf{v}dV \tag{6.3}$$

The rates of linear momentum transport entering and leaving the arbitrary volume (first two terms on the right-hand side of Equation 6.2) can be expressed in terms of the mass flow rates that move across the arbitrary surface. In Chapter 2, we established that the mass flow rate of fluid through an element dA of the arbitrary surface is given by (Equation 2.6)

$$\begin{Bmatrix}\text{Mass of material that}\\ \text{crosses dA per unit time}\end{Bmatrix} = \rho\mathbf{v}_r\cdot\mathbf{n}dA \tag{6.4}$$

where $\mathbf{v}_r$ is the velocity of fluid relative to the arbitrary surface, $\mathbf{v}_r = \mathbf{v} - \mathbf{w}$, with $\mathbf{v}$ and $\mathbf{w}$ being the velocities of the fluid and arbitrary surface, respectively. To find the linear momentum that crosses the surface, we multiply Equation 6.4 by the velocity

$$\begin{Bmatrix}\text{Linear momentum that}\\ \text{crosses dA per unit time}\end{Bmatrix} = \rho\mathbf{v}\mathbf{v}_r\cdot\mathbf{n}dA \tag{6.5}$$

Since $\mathbf{v}_r\cdot\mathbf{n} > \mathbf{0}$ means that the material leaves the arbitrary volume, then the net rate at which linear momentum enters the volume is

$$\begin{Bmatrix}\text{Net linear momentum that}\\ \text{enters the arbitrary volume}\\ \text{through dA per unit time}\end{Bmatrix} = -\rho\,\mathbf{v}(\mathbf{v}-\mathbf{w})\cdot\mathbf{n}dA \tag{6.6}$$

Hence, the total momentum fluxes in Equation 6.2 can be expressed as

$$\begin{Bmatrix}\text{Rate at which linear momentum}\\ \text{enters the arbitrary volume}\end{Bmatrix} - \begin{Bmatrix}\text{Rate at which linear momentum}\\ \text{exits the arbitrary volume}\end{Bmatrix} = \int_{A_a(t)} \rho\,\mathbf{v}(\mathbf{w}-\mathbf{v})\cdot\mathbf{n}dA \tag{6.7}$$

To quantify the forces in the momentum balance (6.2), we first establish that forces acting on a portion of fluid can be of two types:

1. Body or volume forces: these are forces that act on every point of the fluid and are due to external force fields. The most common example is the gravitational force.
2. Surface or contact forces: these are forces exerted on the fluid by its immediate surroundings. They act on the surface of the control volume being considered.

Since forces are additive, we can state:

$$\left\{\begin{matrix}\text{Total force exerted by the} \\ \text{surroundings on a control volume}\end{matrix}\right\} = \left\{\begin{matrix}\text{Body forces exerted by the} \\ \text{surroundings on the control volume}\end{matrix}\right\} + \left\{\begin{matrix}\text{Surface forces exerted by the} \\ \text{surroundings on the control volume}\end{matrix}\right\} \tag{6.8}$$

Here, we will consider the gravitational force as the only body force. The force is then the weight of the fluid: mass × acceleration of gravity. For an element of fluid dV, the body force is $\rho\mathbf{g}dV$. Therefore,

$$\left\{\begin{matrix}\text{Body forces exerted by the} \\ \text{surroundings on the control volume}\end{matrix}\right\} = \int_{V_a(t)} \rho\mathbf{g}dV \tag{6.9}$$

Surface forces are quantified by the stress vector, **t**, defined as the surface force per unit area exerted by the surroundings on the arbitrary volume on each point of the arbitrary surface. The total surface force is given by

$$\left\{\begin{matrix}\text{Surface forces exerted by the} \\ \text{surroundings on the control volume}\end{matrix}\right\} = \int_{A_a(t)} \mathbf{t}dA \tag{6.10}$$

The linear momentum balance 6.2 becomes

$$\frac{d}{dt}\int_{V_a(t)} \rho\mathbf{v}dV = \int_{A_a(t)} \rho\mathbf{v}(\mathbf{w}-\mathbf{v})\cdot\mathbf{n}dA + \int_{V_a(t)} \rho\mathbf{g}dV + \int_{A_a(t)} \mathbf{t}dA \tag{6.11}$$

In the next section we will show how to use this type of balance, in combination with mass balances, to determine velocity distributions in one-dimensional flows.

6.3 VELOCITY DISTRIBUTIONS IN ONE-DIMENSIONAL FLOWS

To illustrate the application of momentum balances, we will consider in this section three case studies of one-dimensional flows.

6.3.1 Velocity Distribution between Parallel Plates When the Fluid Movement Is Induced by the Motion of One of the Plates (Simple Shear Flow)

This flow is a simplification of a classical problem in fluid mechanics: Couette flow. One of the most widely used techniques to measure viscosity of fluids (mainly liquids) is the Couette viscometer, which consists of two concentric cylinders with the fluid occupying the annular space between them (Figure 6.1). One of the cylinders is rotated at a constant angular velocity (ω) while the other is kept stationary. For the case depicted in Figure 6.1, the outer cylinder is rotated. Because of frictional forces, the fluid will try to make the inner cylinder rotate in the same direction as the outer cylinder. Therefore, a torque must be applied to the inner cylinder to keep it from rotating. Here, we will simplify the analysis of this problem by considering that the gap between cylinders is much smaller than the inner cylinder radius, that is, $h \ll R$. Under this condition, the local flow field in the annular space can be approximated to the flow between flat parallel plates separated by a distance h, with one plate stationary and the other moving at a linear velocity $v_0 = \omega(R + h) \approx \omega R$ (Figure 6.2). Note that the motion of the fluid between the plates is driven by the motion of the upper plate; that is, if $v_0 = 0$, the fluid is completely stationary.

First, we need to establish the conditions for which a solution for the specified problem will be sought. The first few conditions are (more will be specified below)

1. We will assume that this is a steady-state flow; that is, the parameters involved in the description of fluid motion do not change with time. This implies that there will be no accumulation terms in mass and momentum balances.
2. Using the coordinate system specified in Figure 6.2, we will assume that the velocity field is independent of y. This is consistent with the fact that

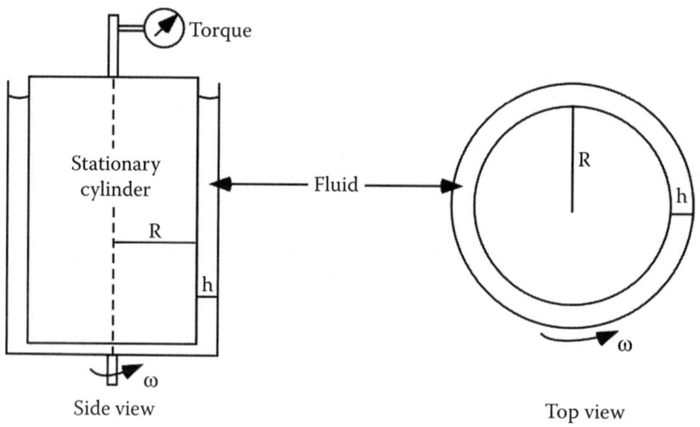

FIGURE 6.1 The Couette viscometer. A fluid occupies the annular space between two concentric cylinders. The outer cylinder is rotated at constant angular velocity (ω) while the inner cylinder is forced to remain stationary by the application of a torque. A measurement of the applied torque is used to determine the fluid's viscosity.

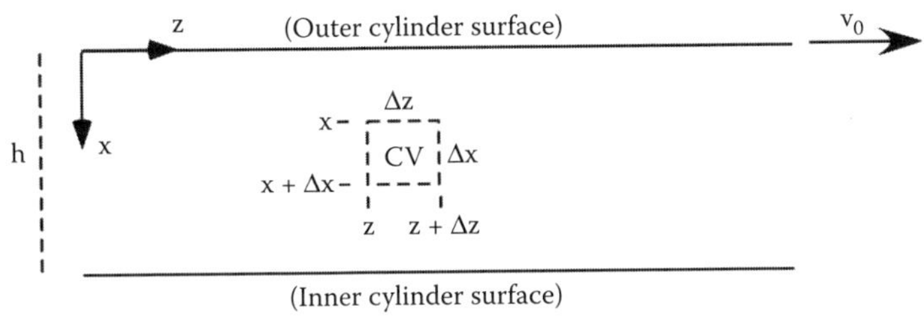

FIGURE 6.2 Fluid motion between parallel plates induced by the motion of the top plate. The control volume (CV) has dimensions $\Delta x \times \Delta z$ and extends a length W in the y direction.

the cause of motion (the motion of the wall) is the same for all values of y. Referring to the Couette geometry (Figure 6.1), this implies that our solution will break down in the vicinity of both top and bottom ends of the apparatus.

3. We will assume that the fluid only moves in the z direction. Due to the y-symmetry of the flow assumed above, we would not expect any motion in y. Additionally, there are no apparent physical causes to generate a motion in the x direction, plus the presence of the two plates prevents motion in the x direction in their vicinity.

In view of these conditions, we expect the velocity field of the fluid between the plates to obey the following:

$$v_x = 0 \tag{6.12}$$

$$v_y = 0 \tag{6.13}$$

$$v_z = v_z(x, z) \tag{6.14}$$

More information on v_z will be obtained by performing a mass balance. For this purpose, consider the control volume shown in Figure 6.2. As studied in Chapter 2, the principle of conservation of mass applied to a control volume can be stated as follows:

$$\left\{\begin{matrix}\text{Time rate of change}\\ \text{of the total mass in}\\ \text{the control volume}\end{matrix}\right\} = \left\{\begin{matrix}\text{Rate at which}\\ \text{mass enters the}\\ \text{control volume}\end{matrix}\right\} - \left\{\begin{matrix}\text{Rate at which}\\ \text{mass leaves the}\\ \text{control volume}\end{matrix}\right\} \tag{6.15}$$

In this case, since the system is at steady state, the left-hand side of this equation is zero. Also, since the fluid is moving in the positive z direction, mass enters the CV at the surface z and leaves at the surface $z + \Delta z$ (Figure 6.2). The mass per unit time crossing each surface (mass flow rate) can be quantified as the product of the fluid density (ρ), its velocity normal to the surface (v_z), and the area of the surface. Hence, the mathematical statement of expression (6.15) is

$$\rho v_z\big|_z W\Delta x = \rho v_z\big|_{z+\Delta z} W\Delta x \tag{6.16}$$

We now impose an additional condition:

4. The density of the fluid is uniform. Flows that satisfy this condition are called **incompressible flows**.

Using this condition, Equation 6.16 simplifies to

$$v_z\big|_z = v_z\big|_{z+\Delta z} \tag{6.17}$$

Since the control volume's location and length are arbitrary (i.e., Δz can be any length in the z direction), this equation implies that v_z does not change with z; that is,

$$v_z = v_z(x) \tag{6.18}$$

Our objective is to find this function, which is termed **velocity profile**. To do this, we will use the principle of conservation of linear momentum. We will write a linear momentum balance for the control volume but first, since linear momentum is a vector, we must specify the direction for the balance as the direction of motion (z). The principle of conservation of linear momentum in the z direction applied to the control volume can be stated as (Equation 6.2)

$$\begin{Bmatrix}\text{Time rate of change of}\\ \text{the total linear momentum}\\ \text{(z direction) in}\\ \text{the control volume}\end{Bmatrix} = \begin{Bmatrix}\text{Rate at which linear}\\ \text{momentum (z direction)}\\ \text{enters the control volume}\end{Bmatrix} - \begin{Bmatrix}\text{Rate at which linear}\\ \text{momentum (z direction)}\\ \text{leaves the control volume}\end{Bmatrix} + \begin{Bmatrix}\text{Total force exerted by the surroundings}\\ \text{on the control volume in the z direction}\end{Bmatrix} \tag{6.19}$$

In this case, the left-hand side of this equation vanishes (steady state). To quantify the two first terms on the right-hand side, we will apply Equation 6.7 to the inlet and outlet cross sections,

$$\begin{Bmatrix}\text{Rate at which linear}\\ \text{momentum (z direction)}\\ \text{enters the control volume}\end{Bmatrix} = \rho W\Delta x\, v_z^2\big|_z \tag{6.20}$$

$$\left\{\begin{matrix}\text{Rate at which linear} \\ \text{momentum (z direction)} \\ \text{leaves the control volume}\end{matrix}\right\} = \rho W \Delta x \, v v_z^2 \big|_{z+\Delta z} \tag{6.21}$$

Since the velocity is independent of z, these two terms are equal and they will cancel upon substitution into Equation 6.19, which then simplifies to

$$\left\{\begin{matrix}\text{Total force exerted by the surroundings} \\ \text{on the control volume in the z direction}\end{matrix}\right\} = 0 \tag{6.22}$$

In other words, since there is neither accumulation of momentum nor net momentum flow in and out of the control volume, the principle of conservation of linear momentum is reduced to a force balance. We now identify the forces acting on the control volume, starting with the body force,

$$\left\{\begin{matrix}\text{Body forces exerted on} \\ \text{the CV in the z direction}\end{matrix}\right\} = \rho \Delta x \Delta z W g_z \tag{6.23}$$

where g_z is the z-component of the acceleration of gravity vector. We will consider that the vertical direction is direction y (Figures 6.1 and 6.2), which implies $g_z = 0$. Therefore, body forces will not appear in our balance. Thus, Equation 6.22 simplifies to

$$\left\{\begin{matrix}\text{Total surface forces exerted by the surroundings} \\ \text{on the control volume in the z direction}\end{matrix}\right\} = 0 \tag{6.24}$$

The control volume selected is delimited by six surfaces. The two surfaces parallel to the x–z plane will be disregarded since there will be no surface force in the z direction there. In this case, we just have to quantify surface forces on the four surfaces shown in Figure 6.3. When we consider the forces in the z direction, we see

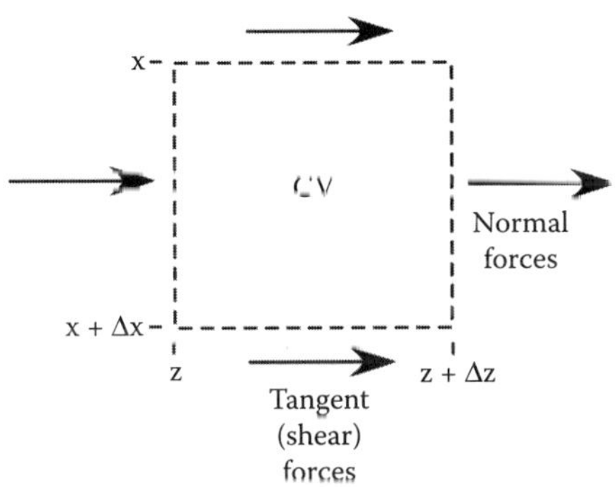

FIGURE 6.3 Surface forces acting on the control volume shown in Figure 6.2.

that two of the forces will be normal to the surface and two forces will be tangent to the surface. Tangent forces will be called **shear forces**.

The surface forces acting on the four surfaces in Figure 6.3 are exerted on the CV **by the fluid surrounding it**. These surface forces will be quantified in terms of the stress, defined as the force per unit area acting on a surface. Normal stresses (i.e., stresses associated with normal forces) are generally equal to the fluid's pressure (P), which is a compression force (i.e., it always acts toward the control volume). On the other hand, the shear stress will be defined as follows:

> τ_{xz}: surface force per unit area exerted **by** the fluid in the CV **on** its surroundings in the z direction on a surface perpendicular to the x axis, with the x axis pointing outward from the CV.

Note that this definition relies on the convention that the stress quantifies forces exerted by the fluid in the CV (as opposed to forces exerted on the fluid). If the opposite convention were used, the sign of the stress would have to be changed, according to the principle of action and reaction. The first index in the symbol for the shear stress will always refer to the direction normal to the surface, whereas the second index will refer to the direction of the force.

Referring to Figure 6.3, on the surface at x, the force that we need is force exerted **on** the fluid, so we would have to use the negative of the shear stress, but, since the x axis points inward to the CV, action and reaction require a second change of sign and the force will be $\tau_{xz}W\Delta z$. On the other hand, on the surface at $x + \Delta x$, the x axis points outwards, but the force required is that acting on the fluid, which would be $-\tau_{xz}W\Delta z$. Figure 6.4 shows the quantification of the forces.

Now, we can express Equation 6.24 in mathematical form as follows:

$$P\big|_{z} W\Delta x - P\big|_{z+\Delta z} W\Delta x + \tau_{xz}\big|_{x} W\Delta z - \tau_{xz}\big|_{x+\Delta x} W\Delta z = 0 \qquad (6.25)$$

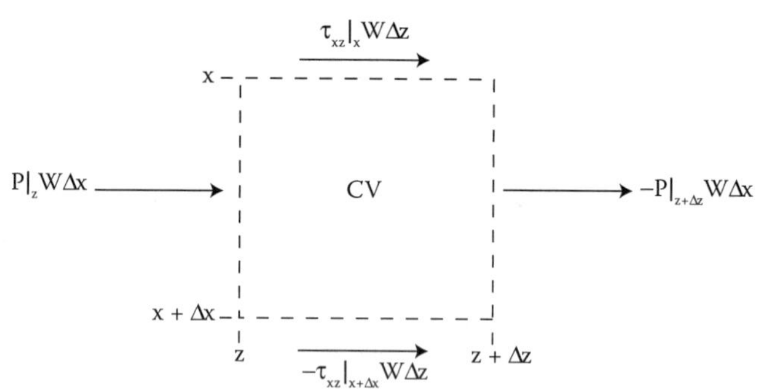

FIGURE 6.4 Quantification of the surface forces in Figure 6.3. The pressure force on the right points into the CV (compression force), which is equivalent to a negative force pointing in the z direction.

Since the fluid motion is only induced by the motion of the upper plate, we would not expect pressure variations along the flow direction (a pressure drop with z will directly "push" the fluid in the direction of motion). Therefore, we have

$$P\Big|_{z} = P\Big|_{z+\Delta z} \tag{6.26}$$

and pressure terms drop out of Equation 6.25, simplifying it to

$$\tau_{xz}\Big|_{x} = \tau_{xz}\Big|_{x+\Delta x} \tag{6.27}$$

The momentum balance leads directly to the consequence that the shear stress is independent of x. Dividing Equation 6.27 by Δx and taking the limit as $\Delta x \to 0$ leads to

$$\frac{d\tau_{xz}}{dx} = 0 \tag{6.28}$$

Or, simply stated, the shear stress is a constant in the flow field. The shear stress is directly related to the motion of the fluid. Note that, at $x = 0$ (upper wall), the shear stress is a consequence of the wall exerting a shear force on the fluid. The fact that τ_{xz} is independent of x means that the stress exerted by the wall on the fluid is propagated throughout the flow field. These forces are propagated by the interaction between fluid molecules from different fluid layers in x. At $x = 0$, the interaction is between wall and fluid molecules. The molecular-level interaction can be viewed as a tendency of fluid molecules to "stick" to each other (or to the wall's molecules at $x = 0$), so the shear stress is what is known as a frictional force. Friction would not exist between two adjacent fluid layers unless they were moving at different velocities. Therefore, the existence of a shear stress is related to **changes** in v_z with x. In other words, we would not expect τ_{xz} to depend directly on the fluid velocity but on the velocity gradient, dv_z/dx. It is well known that the shear stress is directly proportional to the velocity gradient for a large number of practical cases. Fluids that obey this type of relation are called **Newtonian fluids**. Hence, for a Newtonian fluid,

$$\tau_{xz} \propto \frac{dv_z}{dx} \tag{6.29}$$

As shown in Figure 1.4, and according to the convention established in the definition of τ_{xz}, one would expect $\tau_{xz} > 0$, since the moving plate exerts a friction force on the fluid in the positive z direction. On the other hand, the fluid velocity should decrease with x, since it goes from the velocity of the plate,

$$v_z = v_0 \text{ at } x = 0 \tag{6.30}$$

to

$$v_z = 0 \text{ at } x = h \tag{6.31}$$

Here, it is assumed that the fluid sticks to the solid walls, which occurs in a majority of applications. For this reason, Equations 6.30 and 6.31 are called **no-slip conditions**. The fact that the velocity decreases with x means that $dv_z/dx < 0$ and, therefore, the proportionality relation 6.28 must include a minus sign. Formally, the relation between shear stress and velocity gradient in this flow is written as (e.g., Bird et al. 2007)

$$\tau_{xz} = -\mu \frac{dv_z}{dx}, \text{ for a Newtonian fluid} \tag{6.32}$$

where the proportionality constant μ is the fluid viscosity. It is clear that Equation 6.32 is a macroscopic consequence of molecular effects, and that the viscosity reflects the effect that molecular interactions have on frictional forces.

Table 6.1 shows values of density and viscosity for some common gases and liquids. It is evident from these values that both properties (even though they are both consequences of molecular level effects) are independent of each other. In addition, the order of magnitude of liquid viscosities varies widely. All fluids listed in Table 6.1 are Newtonian. The viscosity depends on the molecular structure of the substance

TABLE 6.1
Densities and Viscosities of Some Fluids

Substance	ρ (g/cm³)	μ (mPa s)
Gases		
Hydrogen	0.083×10^{-3}	0.0086
Methane	0.666×10^{-3}	0.0109
Oxygen	1.33×10^{-3}	0.0204
Air	1.2×10^{-3}	0.018
Liquids		
Water	1.00	1.003
Water (30°C)	1.00	0.798
Ethanol	0.77	1.216
Glycerine	1.26	~1000
Motor oil SAE 30	0.89	350
Mercury	13.6	1.55
Honey	1.2–1.6	2000–10,000
Soda glass (570°C)[a]	2.4	$\sim 10^8$
Supercooled He[b]	–	0

Note: Properties are listed at 1 atm, 20°C unless otherwise noted.

[a] Softening point of the glass: temperature at which the glass sags appreciably in a period of hours.

[b] Temperatures below 2K.

and also on its thermodynamic state. For a pure substance, the viscosity is more sensitive to temperature than to pressure. Note how drastic the viscosity changes with temperature is for liquid water.

Now that we have a representation for the shear stress in terms of the velocity field, we can substitute it into the momentum balance (Equation 6.28) to get

$$\frac{d}{dx}\left(\mu\frac{dv_z}{dx}\right) = 0 \tag{6.33}$$

We now impose an additional condition to the flow:

5. The viscosity is uniform (this will be strictly true if the pressure, temperature, and composition of the fluid are uniform, but moderate changes would make this a good approximation in some cases).

Equation 6.33 then becomes

$$\frac{d^2v_z}{dx^2} = 0 \tag{6.34}$$

This equation can be integrated twice to find

$$v_z = C_1x + C_2 \tag{6.35}$$

where C_1 and C_2 are integration constants, which can be evaluated from the boundary conditions 6.30 and 6.31 to get

$$v_z = v_0\left(1 - \frac{x}{h}\right) \tag{6.36}$$

This equation is the velocity profile for the fluid between the plates. The shear stress can be evaluated now directly from Equation 6.32 to obtain

$$\tau_{xz} = \frac{\mu v_0}{h} \tag{6.37}$$

The velocity and shear stress profiles are shown in Figure 6.5.

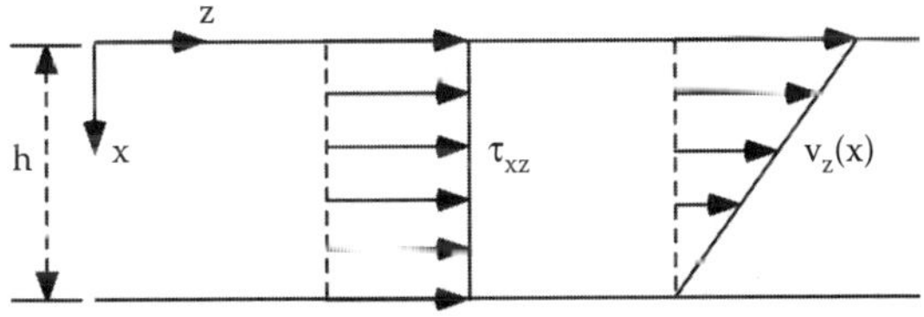

FIGURE 6.5 Velocity and shear stress profiles for flow between parallel plates. The shear stress is uniform and the velocity varies linearly with position.

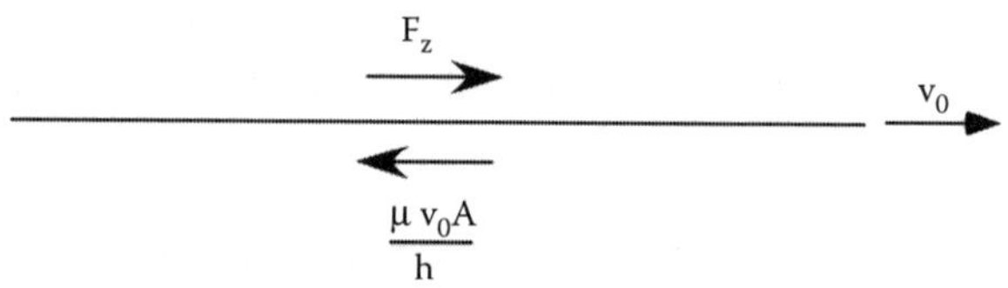

FIGURE 6.6 Force balance on the upper plate.

If A is the surface area of the upper plate, the force exerted by the upper plate on the fluid is the product of the shear stress times A,

$$\tau_{xz} A = \frac{\mu v_0 A}{h} \tag{6.38}$$

By action and reaction, the fluid exerts the same force but in opposite direction on the plate. Since the plate is moving at a constant velocity, the sum of all forces acting on it must be zero. If F_z is the force that must be applied to the plate, then (Figure 6.6)

$$F_z = \frac{\mu v_0 A}{h} \tag{6.39}$$

It is clear that the driving force for motion in this problem is the motion of the upper plate, which is a consequence of this force. This type of flow is called **simple shear flow** on the basis of the type of deformation experienced by the fluid.

6.3.2 Velocity Distribution in a Liquid Film Flowing Down on an Inclined Surface

Consider a thin film of liquid flowing over a large, flat, inclined surface under the effect of gravity. The geometry of the system is shown in Figure 6.7.

Assumptions and conditions for this case are

1. Steady state (includes the absence of waves or instabilities on the free surface).
2. Incompressible flow (constant density).
3. The flow is one-dimensional: $v_x \neq 0$, $v_y = v_z = 0$. This implies that the film thickness is uniform. This restricts the analysis to regions that are far from the film formation.
4. Flow parameters are independent of z.
5. The liquid is Newtonian and has a uniform viscosity.

The analysis will be similar to the previous problem with two main differences: first, in this case the driving force for the flow is gravity, and second, there is a free surface through which the liquid contacts another fluid phase (air). Our aim is to find the velocity field, which, assuming symmetry in the z direction, has the form $v_x(x,y)$.

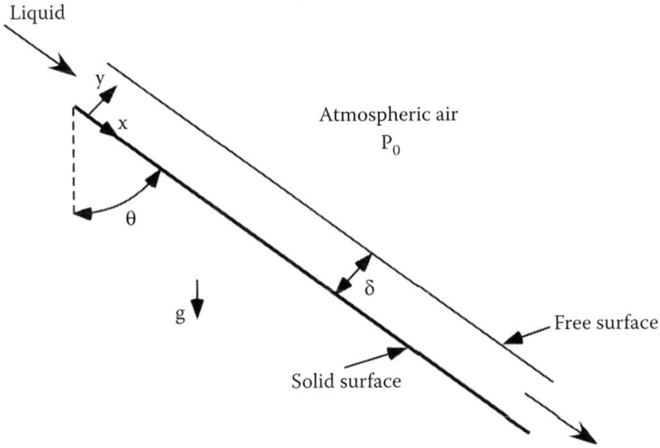

FIGURE 6.7 A liquid film of uniform thickness δ flows down an inclined surface. The free surface is the liquid/air interface. The motion of the liquid is due exclusively to the gravitational pull.

To analyze this problem, we select the control volume shown in Figure 6.8. We start by applying the mass balance, Equation 6.15. Since the process is at steady state, we only need to consider input and output terms:

$$\rho v_x\big|_x W\Delta y = \rho v_x\big|_{x+\Delta x} W\Delta y \tag{6.40}$$

Since the density is a constant, this equation yields

$$v_x\big|_x = v_x\big|_{x+\Delta x} \tag{6.41}$$

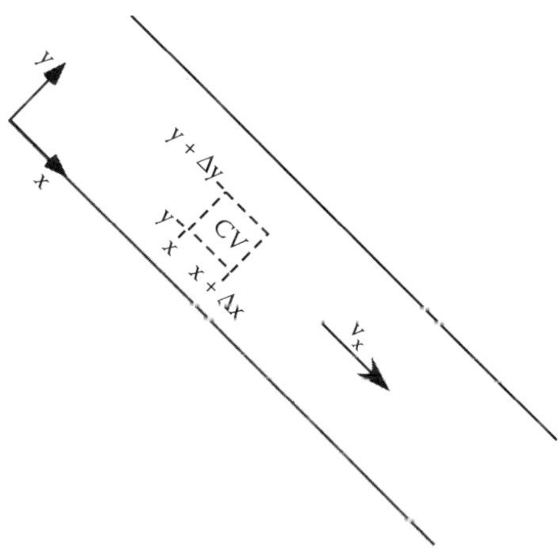

FIGURE 6.8 Control volume (CV) for the analysis of film flows down an inclined surface. The CV has dimensions $\Delta x \times \Delta y$ and extends a length W in the z direction.

Once again, mass conservation implies that the velocity does not change in the direction of motion, which means that $v_x = v_x(y)$.

Now, we apply conservation of linear momentum in the x direction. As was the case in the previous section, the fact that the velocity does not change with the direction of motion implies that momentum fluxes in and out of the CV are the same and the momentum balance is reduced to a force balance (similar to Equation 6.22). Figure 6.9 shows the forces exerted on the CV acting in the x direction. In this case, the projection of the gravitational force in the x direction needs to be considered in the balance. This force is equal to the weight of fluid in the CV, $M = \rho\Delta V$, times the acceleration of gravity in the x direction, $g_x = g\cos\theta$ (Figure 6.9). The force balance in the x direction is

$$P\Big|_x W\Delta y + \tau_{yx}\Big|_y W\Delta x + \rho W\Delta x\Delta y g\cos\theta = P\Big|_{x+\Delta x} W\Delta y + \tau_{yx}\Big|_{y+\Delta y} W\Delta x \quad (6.42)$$

The free surface of the liquid is exposed to atmospheric pressure at all values of x. Therefore, we expect that

$$P\Big|_x = P\Big|_{x+\Delta x} \quad (6.43)$$

which eliminates the pressure terms in Equation 6.42. Dividing the resulting equation by $W\Delta x\Delta y$ and rearranging leads to

$$\frac{\tau_{yx}\Big|_{y+\Delta y} - \tau_{yx}\Big|_y}{\Delta y} = \rho g\cos\theta \quad (6.44)$$

Taking the limit as $\Delta y \to 0$ yields

$$\frac{d\tau_{yx}}{dy} = \rho g\cos\theta \quad (6.45)$$

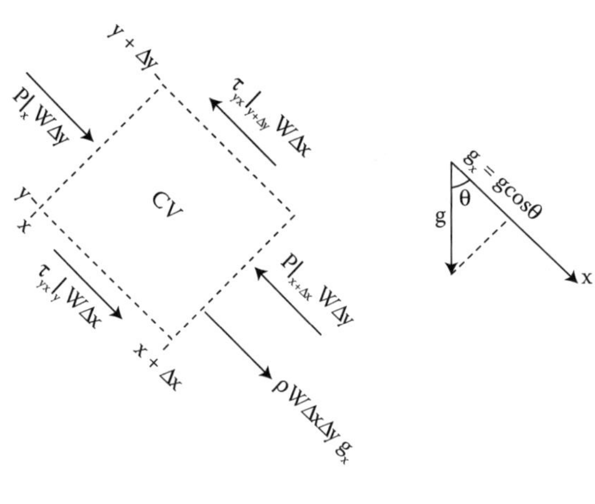

FIGURE 6.9 Forces acting on the CV in Figure 6.8.

Since the right-hand side of this equation is a constant, it can be integrated to obtain

$$\tau_{yx} = y\rho g\cos\theta + C_1 \tag{6.46}$$

where C_1 is an integration constant. At the free surface, the liquid is in contact with atmospheric air. Any shear force applied on this surface on the liquid would have to be applied by the air. However, air has a relatively low viscosity (at 20°C and 1 atm, $\mu_{air}/\mu_{water} = 0.02$, Table 6.1) and, for this reason, it is not likely to exert an appreciable shear force on the liquid; that is,

$$\tau_{yx} \approx 0, \text{ at } y = \delta(\text{free surface}) \tag{6.47}$$

Using this equation as a boundary condition in Equation 6.46, we can find C_1, and subsequent substitution into Equation 6.46 yields the shear stress profile

$$\tau_{yx} = -(\delta - y)\rho g\cos\theta \tag{6.48}$$

Note that $\tau_{yx} \leq 0$. The constitutive equation for a Newtonian fluid in this case is

$$\tau_{yx} = -\mu\frac{dv_x}{dy} \tag{6.49}$$

Substitution into Equation 6.48 leads to

$$\frac{dv_x}{dy} = (\delta - y)\frac{\rho g\cos\theta}{\mu} \tag{6.50}$$

Integrating with respect to y yields

$$v_x = \left(\delta y - \frac{y^2}{2}\right)\frac{\rho g\cos\theta}{\mu} + C_2 \tag{6.51}$$

The integration constant C_2 can be found using the no-slip condition

$$v_x = 0, y = 0 \tag{6.52}$$

which yields $C_2 = 0$, and the velocity profile is

$$v_x = y\left(\delta - \frac{y}{2}\right)\frac{\rho g\cos\theta}{\mu} \tag{6.53}$$

The velocity and shear stress profiles are shown in Figure 6.10. The velocity profile is parabolic with an absolute maximum at the free surface (note that Equation 6.50 implies that the velocity derivative at the free surface is zero). The maximum velocity is v_x at $y = \delta$,

$$v_{x,max} = \frac{\rho g \cos\theta \delta^2}{2\mu} \tag{6.54}$$

It is clear from this equation that the gravitational force is responsible for the motion of the liquid. For a horizontal plane ($\theta = \pi/2$), we would have: $v_x \equiv 0$. On the other hand, the maximum shear stress (in magnitude) occurs at the surface of the plane ($y = 0$) for which (Equation 6.48)

$$\tau_{yx}\Big|_{y=0} = -\delta\rho g \cos\theta \tag{6.55}$$

This shear stress reflects the force that the fluid exerts on the surface (F_x). According to our convention, since the $-y$ axis points away from the CV at $y = 0$, we have

$$F_x = -A\tau_{yx}\Big|_{y=0} = A\delta\rho g \cos\theta \tag{6.56}$$

where A is the surface area of the plane covered by the liquid film.

The volumetric flow rate of liquid in the film (Q) is obtained by integrating the velocity profile over the cross-sectional area (A_c),

$$Q = \int_{A_c} v_x dA_c \tag{6.57}$$

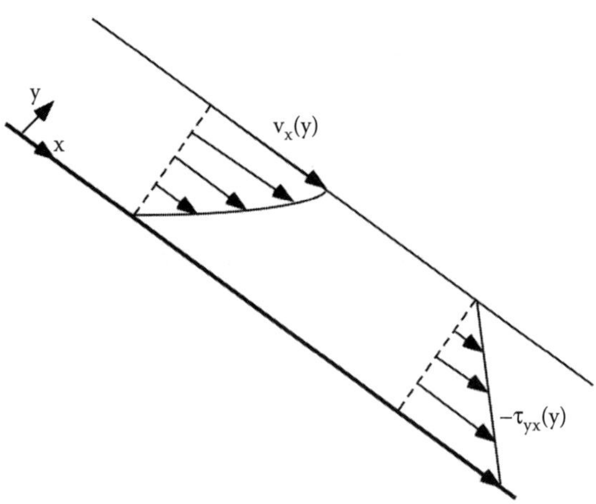

FIGURE 6.10 Velocity and shear stress profiles for flow of a liquid film down an inclined plate.

In this case, the cross section is a rectangle in the y–z plane of dimensions $\delta \times W$, so that

$$Q = \int_0^W \int_0^\delta v_x dydz = W \int_0^\delta v_x dy \tag{6.58}$$

Substituting the velocity profile (Equation 6.53) and integrating yields, after manipulations,

$$Q = \frac{W\rho g \cos\theta \delta^3}{3\mu} \tag{6.59}$$

Note the sensitivity of the flow rate to the film thickness: to double the film thickness, the flow rate would have to increase eightfold. The average velocity is the ratio between the flow rate and the cross-sectional area,

$$<v_x> = \frac{Q}{A_c} = \frac{Q}{W\delta} = \frac{\rho g \cos\theta \delta^2}{3\mu} \tag{6.60}$$

Note that (see Equation 6.54)

$$<v_x> = \frac{2}{3} v_{x,max} \tag{6.61}$$

One of the most important assumptions in the solution found is the fact that the flow is one dimensional and, consequently, the film has a uniform thickness. In practical situations, this will hold only for relatively thin films, which will be achieved at relatively low flow rates. If the flow rate is increased on a stable, uniform film, instabilities in the flow eventually appear which leads to surface waves and eventually even to a film breakup. As an example, consider water flowing down a 45° slope. Just to have an idea of the order of magnitude of stable film thicknesses that can be achieved in this case, consider that the surface of the film is moving relatively fast: $v_x\big|_{x=\delta} = v_{x,max} = 1\,\text{m/s}$. For water at 20°C, $\mu = 10^{-3}$ Pa s, $\rho = 1000$ kg/m^3. Using $g = 9.81$ m/s, we can calculate the film thickness from Equation 6.54: $\delta = 0.54$ mm.

6.3.3 Velocity Distribution for Developed Flow in a Pipe

As a final example of the application of mass and momentum conservation to flow problems, consider vertical flow in a pipe with circular cross section (Figure 6.11). The pipe has radius R and through the length of pipe L, the fluid pressure changes from P_0 to P_L. The problem will be analyzed using cylindrical coordinates.

Assumptions and conditions for this problem are

1. Steady state.
2. Incompressible flow.

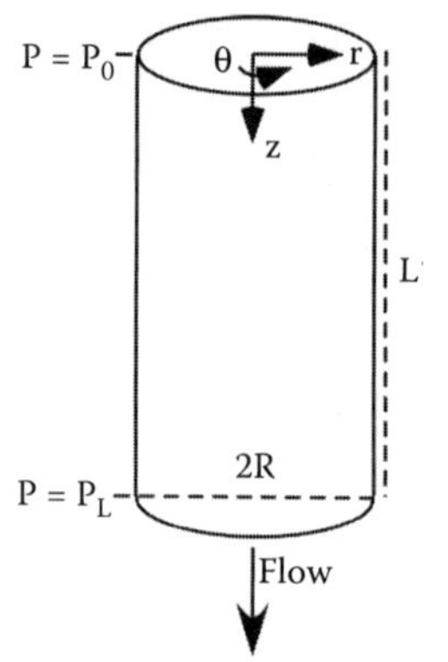

FIGURE 6.11 Flow of a liquid inside a cylindrical pipe.

3. The flow is one-dimensional in the z direction. This condition implies that the flow is developed (far from entrances and exits or changes in geometry).
4. Angular symmetry (flow parameters are independent of the angular coordinate θ).
5. The liquid is Newtonian and has a uniform viscosity.

Accordingly, the components of the velocity vector are: $v_r = 0$, $v_\theta = 0$, $v_z = v_z(r,z)$.

To analyze this problem, we will use the control volume shown in Figure 6.12, which is spanned by small but finite changes in the three coordinates. The dimensions of the control volume are $\Delta r \times r\Delta\theta \times \Delta z$.

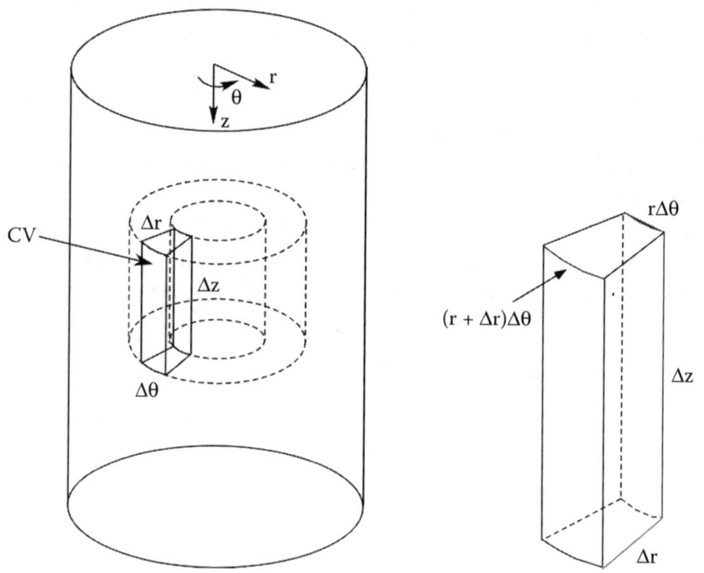

FIGURE 6.12 Three-dimensional elementary control volume in cylindrical coordinates (left) and lengths of each side (right).

The control volume is delimited by six surfaces, but we notice that mass of fluid only crosses the top and bottom surfaces, since the only nonzero component of the velocity vector is v_z. Figure 6.13 shows the mass flow rates entering and leaving the control volume. According to mass conservation, these flow rates are the same:

$$\rho v_z\big|_z r\Delta\theta\Delta r = \rho v_z\big|_{z+\Delta z} r\Delta\theta\Delta r \tag{6.62}$$

which implies that

$$v_z\big|_z = v_z\big|_{z+\Delta z} \tag{6.63}$$

that is, v_z is independent of z and, therefore, $v_z = v_z(r)$.

To apply momentum conservation in the z direction, we again recognize that the rates of flow of linear momentum entering and leaving the control volume are the same, and the momentum balance reduces to a force balance. In this case, we have to consider both surface and body forces. The surface forces in the z direction on the top and bottom surfaces will be due to normal stresses, considered to be equal to the fluid pressure.

On the other four surfaces, the surface forces in z are due to shear stresses. On the surfaces at θ and θ + Δθ, we do not expect to have shear stresses because of symmetry (note that v_z does not change with θ). Figure 6.14 shows the forces in the z direction acting on the control volume. Note that the areas of the lateral surfaces are not the same due to changes in the radial position (see Figure 6.12).

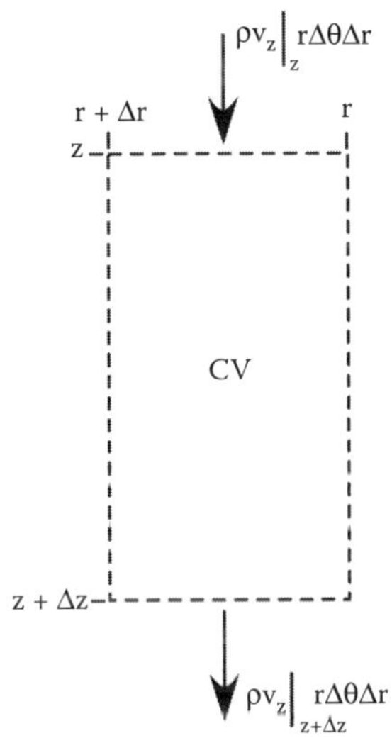

FIGURE 6.13 Side view of the control volume in Figure 6.12, showing mass flow rates crossing its surfaces.

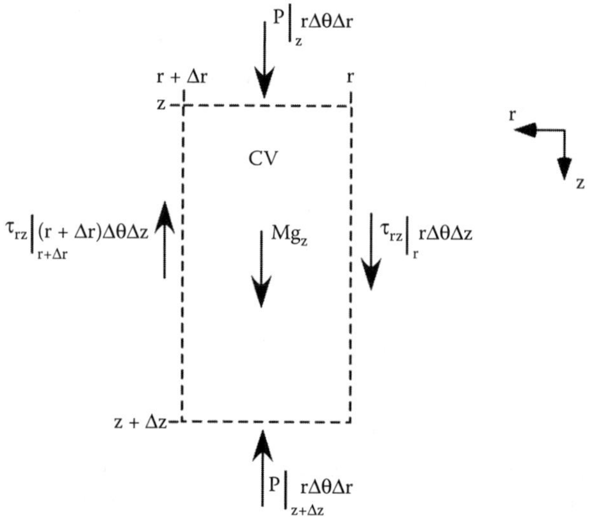

FIGURE 6.14 Forces in the z direction acting on the control volume.

The mass of fluid in the control volume is

$$M = \rho \Delta r \, r\Delta\theta\Delta z \tag{6.64}$$

The force balance in the z direction is

$$P\big|_z \, r\Delta\theta\Delta r + \tau_{rz}\big|_r \, r\Delta\theta\Delta z = P\big|_{z+\Delta z} \, r\Delta\theta\Delta r + \tau_{rz}\big|_{r+\Delta r}(r+\Delta r)\Delta\theta\Delta z - \rho g_z r\Delta\theta\Delta r\Delta z \tag{6.65}$$

The shear stress term at r + Δr can be expressed as

$$\tau_{rz}\big|_{r+\Delta r}(r+\Delta r)\Delta\theta\Delta z = (r\tau_{rz})\big|_{r+\Delta r}\Delta\theta\Delta z \tag{6.66}$$

Substituting this equation into Equation 6.65 and dividing by $\Delta r\Delta\theta\Delta z$ yields, after rearranging,

$$-r\frac{P\big|_{z+\Delta z} - P\big|_z}{\Delta z} + \rho g_z r - \frac{(r\tau_{rz})\big|_{r+\Delta r} - (r\tau_{rz})\big|_r}{\Delta r} = 0 \tag{6.67}$$

We now let $\Delta r, \Delta z \to 0$ to obtain

$$-r\frac{\partial P}{\partial z} + \rho g_z r - \frac{\partial (r\tau_{rz})}{\partial r} = 0 \tag{6.68}$$

Since there is no motion in the r direction, a momentum balance in r can be used to show that P is independent of r. Since P is also independent of θ because of symmetry, we conclude that P = P(z). On the other hand, the shear stress for a Newtonian fluid in this case is given by the constitutive equation

$$\tau_{rz} = -\mu \frac{dv_z}{dr} \tag{6.69}$$

This equation implies that the shear stress is only a function of r, $\tau_{rz} = \tau_{rz}(r)$. The partial derivatives in Equation 6.68 then become total derivatives,

$$-\frac{dP}{dz} + \rho g_z = \frac{1}{r}\frac{d(r\tau_{rz})}{dr} \tag{6.70}$$

This equation seems at first to be a contradiction: the left-hand side is only a function of z while the right-hand side is only a function of r. Since r and z are both independent variables, they cannot be related to each other, as Equation 6.70 seems to indicate. The only way to avoid the contradiction is to recognize that two functions of two mutually independent variables can be equal only if the functions are the constant function. Therefore, we state

$$-\frac{dP}{dz} + \rho g_z = \frac{1}{r}\frac{d(r\tau_{rz})}{dr} = C \tag{6.71}$$

where C is a constant (undefined at this point). This leads to two independent ordinary differential equations,

$$-\frac{dP}{dz} + \rho g_z = C \tag{6.72}$$

$$\frac{1}{r}\frac{d(r\tau_{rz})}{dr} = C \tag{6.73}$$

Since the density and acceleration of gravity are constants, Equation 6.72 can be integrated to obtain

$$P = (\rho g_z - C)z + C_1 \tag{6.74}$$

where C_1 is an integration constant. This equation implies that the pressure varies **linearly** with z. The two constants can be found by applying the conditions (see Figure 6.11)

$$P = P_0, \; z = 0 \tag{6.75}$$

$$P = P_L, z = L \tag{6.76}$$

Equation 6.75 implies $C_1 = P_0$, and C is found from Equation 6.76,

$$C = \frac{\Delta P}{L} + \rho g_z \tag{6.77}$$

where $\Delta P = P_0 - P_L$ is the pressure drop that occurs over a length L of pipe. The pressure profile becomes

$$P = -\frac{\Delta P}{L} z + P_0 \tag{6.78}$$

It is common to combine pressure and gravitational effects by using the modified pressure, which in this case is defined by

$$\mathcal{P} = P - \rho g_z z \tag{6.79}$$

so that

$$\mathcal{P}_0 = P_0 \tag{6.80}$$

$$\mathcal{P}_L = P_L - \rho g_z L \tag{6.81}$$

The modified pressure drop is $\Delta\mathcal{P} = \mathcal{P}_0 - \mathcal{P}_L$; from Equations 6.77, 6.80, and 6.81, we get

$$\frac{\Delta\mathcal{P}}{L} = \frac{\Delta P}{L} + \rho g_z = C \tag{6.82}$$

Equation 6.73 can be expressed as

$$\frac{1}{r}\frac{d(r\tau_{rz})}{dr} = \frac{\Delta\mathcal{P}}{L} \tag{6.83}$$

Substituting the constitutive Equation 6.69 into this equation, we get an ordinary differential equation for the velocity field,

$$\frac{d}{dr}\left(r\frac{dv_z}{dr}\right) = -\frac{r}{\mu}\frac{\Delta\mathcal{P}}{L} \tag{6.84}$$

where we have used the condition that the viscosity is uniform. This equation can be integrated once to obtain

$$\frac{dv_z}{dr} = -\frac{1}{2\mu}\frac{\Delta\mathcal{P}}{L}r + \frac{C_2}{r} \tag{6.85}$$

A second integration leads to

$$v_z = -\frac{1}{4\mu}\frac{\Delta\mathcal{P}}{L}r^2 + C_2 \ln r + C_3 \tag{6.86}$$

To find the integration constants, we need two boundary conditions. One will be the no-slip condition at the pipe wall,

$$v_z = 0, r = R \tag{6.87}$$

The profile in Equation 6.86 has a term that diverges as $r \to 0$. However, a physical condition for this problem would be

$$v_z \text{ is finite}, r = 0 \tag{6.88}$$

which requires that $C_2 = 0$. With this in mind and applying Equation 6.87 in 6.86, we can find C_3, which leads to

$$v_z = \frac{1}{4\mu}\frac{\Delta\mathcal{P}}{L}(R^2 - r^2) \tag{6.89}$$

It is interesting to note that the physical condition 6.87 has the exact same mathematical consequence as any of the following two conditions:

1. Symmetry of the velocity profile around the cylinder axis:

$$\frac{dv_z}{dr} = 0, r = 0 \tag{6.90}$$

2. No shear stress at the cylinder axis:

$$\tau_{rz} = 0, r = 0 \tag{6.91}$$

(Equation 6.91 leads to Equation 6.90 after using the constitutive Equation 6.69.)

The shear stress profile can be obtained directly by substituting Equation 6.89 into Equation 6.69. The result is

$$\tau_{rz} = \frac{\Delta\mathcal{P}}{L}\frac{r}{2} \tag{6.92}$$

The velocity and shear stress profiles are shown in Figure 6.15. As was the case in the previous example, the shear stress profile is linear with position while the velocity profile is parabolic.

The maximum velocity occurs at the cylinder axis (centerline), r = 0:

$$v_{z,max} = \frac{R^2}{4\mu}\frac{\Delta\mathcal{P}}{L} \tag{6.93}$$

While the shear stress is maximum at the wall (r = R):

$$\tau_{rz,max} = \frac{\Delta\mathcal{P}}{L}\frac{R}{2} \tag{6.94}$$

Note that the velocity profile can be expressed in terms of the maximum velocity (using Equation 6.93 in Equation 6.89) as follows:

$$v_z = v_{z,max}\left(1 - \frac{r^2}{R^2}\right) \tag{6.95}$$

The volumetric flow rate can be found by integration of the velocity profile over the cross section:

$$Q = \int_{A_c} v_z dA_c \tag{6.96}$$

In cylindrical coordinates (see Figure 6.16), $dA_c = r\,dr\,d\theta$. The integral in this equation can be expressed as follows:

$$Q = \int_0^{2\pi}\int_0^R v_z r dr d\theta = 2\pi\int_0^R v_z r dr \tag{6.97}$$

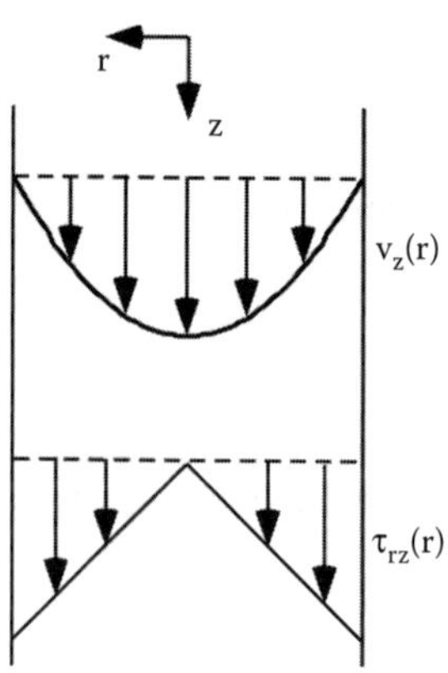

FIGURE 6.15 Velocity and shear stress profiles for steady, developed flow in a pipe.

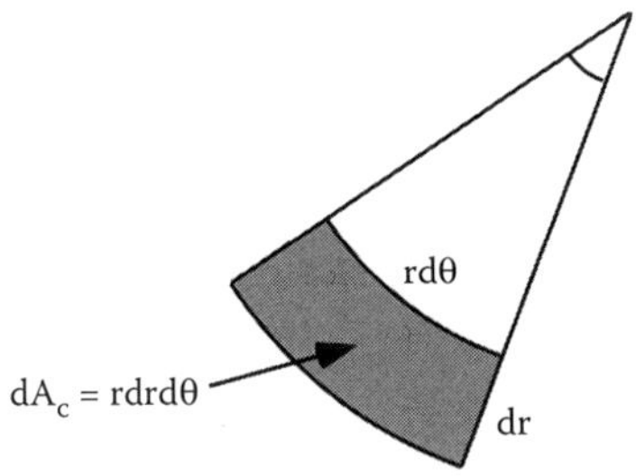

FIGURE 6.16 Differential element of area on the r–θ plane in cylindrical coordinates.

Substituting Equation 6.88 into Equation 6.97 and integrating yields

$$Q = \frac{\pi R^4}{8\mu} \frac{\Delta \mathcal{P}}{L} \tag{6.98}$$

This equation is known as the **Hagen–Poiseuille equation**, and it is one of the most well-known equations in fluid mechanics. Note that the driving force for fluid motion is the modified pressure drop, which contains two contributions: the pressure drop and a gravitational effect (Equation 6.82). For a static fluid (Q = 0), Equation 6.97 implies $\Delta \mathcal{P} = 0$ which, according to Equation 6.82, leads to

$$\frac{\Delta P}{L} = -\rho g_z \tag{6.99}$$

which is consistent with a hydrostatic pressure distribution. Also, if $\Delta P = 0$ (e.g., if both ends of the pipe are open to the atmosphere), then $\Delta \mathcal{P} = \rho g_z L$ (Equation 6.82), which means that the motion of the fluid would be only a result of the gravitational force.

The average velocity is given by

$$\langle v_z \rangle = \frac{Q}{\pi R^2} = \frac{R^2}{8\mu} \frac{\Delta \mathcal{P}}{L} \tag{6.100}$$

Comparison of Equations 6.93 and 6.100 yields

$$\langle v_z \rangle = \frac{v_{z,max}}{2} \tag{6.101}$$

and the velocity profile (Equation 6.95) can be expressed as follows:

$$v_z = 2\langle v_z \rangle \left(1 - \frac{r^2}{R^2}\right) \tag{6.102}$$

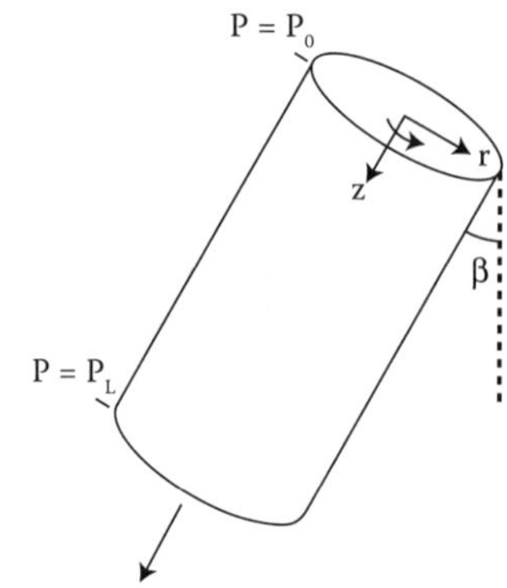

FIGURE 6.17 Coordinate system and geometry for an inclined pipe.

The solution obtained in this section is applicable to steady flows in pipes. For a given configuration, it is well known that there is a velocity over which steady flows cannot be obtained in a pipe. The flow becomes unstable with rapid variations of the velocity with time. This flow regime is called **turbulent flow**. The solution found is only valid when the Reynolds number defined by

$$\mathrm{Re} = \frac{\rho \langle v_z \rangle (2R)}{\mu} \tag{6.103}$$

is small enough (typically Re < 2100 ensures stable flow).

In the above analysis, we have kept the acceleration of gravity component as g_z. In downwards flow, it is evident that $g_z = g$ (Figure 6.11). However, the results presented in this section can be proven to hold even for inclined pipes. For a pipe forming an angle β with the vertical direction (Figure 6.17), the velocity and shear stress profiles, and the Hagen–Poiseuille equation, are still valid, using $g_z = g \cos \beta$, which is the projection of the acceleration of gravity vector in the direction along the pipe (z in Figure 6.17). For a horizontal pipe, $g_z = 0$ and the modified pressure drop is equal to the pressure drop (Equation 6.82), which means that the pressure drop in the pipe is the only driving force for motion.

6.4 CONVECTIVE MASS TRANSFER IN THE ENTRANCE REGION OF A PIPE IN LAMINAR FLOW

As a case study for the treatment of convective mass transport problems using the convection–diffusion equation, consider a section of a pipe that has a soluble solid deposit (species A) on the wall. The deposit will be removed by the flow of a solvent (species B) through the pipe in laminar, developed flow (Figure 6.18). This could correspond, for example, to the cleaning of pipes used to transport milk in the dairy industry. Proteins, bacteria, and precipitated calcium salts tend to deposit on pipe surfaces. In particular, bacteria may grow and pose a risk to human health. For this reason, pipes are cleaned frequently, including the use of solvents that solubilize the deposits.

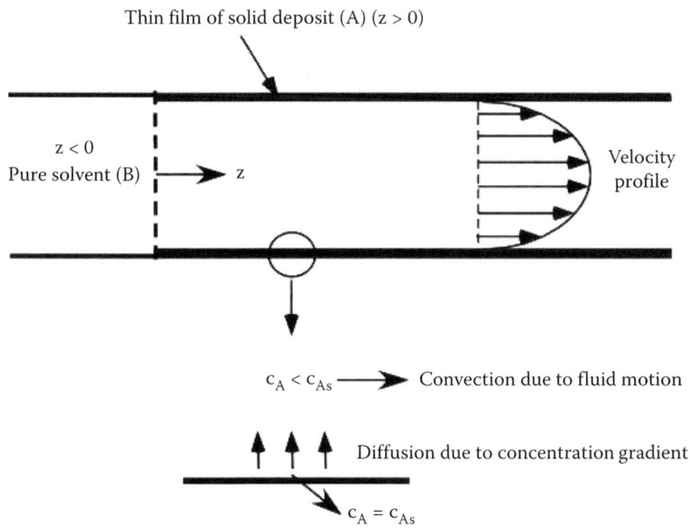

FIGURE 6.18 A thin film of a soluble solid deposit (A) will be cleaned from the internal surface of a pipe (radius R) by the flow of a solvent (B). Here, the pure solvent (B) enters (z = 0) a section of pipe with a thin deposit of solid (A) that is soluble in B. At the pipe surface (r = R), the liquid rapidly equilibrates with the solid deposit, acquiring a concentration equal to the solubility of A in B (c_{As}). This creates a radial concentration gradient so that A diffuses from the pipe surface to the bulk fluid.

The liquid leaving the section of the pipe under analysis will be a mixture of A and B. Species A enters the liquid by dissolution of the film (i.e., A diffuses in the radial direction) whereas B moves in the z direction, dragging A along. Consider what happens in the vicinity of the film surface (Figure 6.18). Since the liquid velocity is zero on the film surface, there will be a thin region adjacent to the surface in which the liquid is moving slowly. In this region, the liquid will achieve equilibrium with the solid (local equilibrium), so that the concentration of A in the liquid will be equal to the solubility of A in B at the conditions of the process (c_{As}), which is the maximum concentration of A that the liquid B can have. Away from the surface, the concentration of A will drop to a low value, since A will be carried away by the liquid flow. A net diffusive flux of A will be established from the film surface to the liquid. The molecules of A that migrate into the flowing liquid will become part of the main flow and will travel downstream with the fluid.

We will assume that the fluid flow is developed, so that the velocity profile is given by Equation 6.102. Furthermore, we will consider that the process is at steady state. The concentration of solute for z > 0 will be a function of both radial and axial positions: $c_A = c_A(r,z)$. Under these conditions, the convective–diffusion equation in cylindrical coordinates (Equation C.12) simplifies to

$$v_z \frac{\partial c_A}{\partial z} = D_{AB}\left[\frac{1}{r}\frac{\partial}{\partial r}\left(r\frac{\partial c_A}{\partial r}\right) + \frac{\partial^2 c_A}{\partial z^2}\right] \tag{6.104}$$

The boundary conditions are

1. It will be assumed that the fluid reaches cleaning section ($z > 0$) free of solute:

$$c_A = 0, z = 0 \tag{6.105}$$

2. For $z > 0$ (solution domain), the concentration of solute in contact with the solid deposit is the solubility,

$$c_A = c_{As}, r = R \tag{6.106}$$

3. The concentration is symmetric around the pipe axis:

$$\frac{\partial c_A}{\partial r} = 0, r = 0 \tag{6.107}$$

At this point, an additional boundary condition in the z direction would be needed to solve the problem, since the differential Equation 6.104 is second order in z. However, we will make an additional simplification that will lead to a formulation in which that additional boundary condition is not needed.

Each term in the resulting differential Equation 6.104 can be identified as a mass transport mechanism:

$$v_z \frac{\partial c_A}{\partial z} \text{ represents the axial (z) transport of species by convection}$$

$$\frac{D_{AB}}{r}\frac{\partial}{\partial r}\left(r\frac{\partial c_A}{\partial r}\right) \text{ represents diffusion in the radial direction}$$

Note that this mechanism will be directly responsible for the concentration change in the fluid, since it will account for species mass transfer the wall (Figure 6.18).

$$D_{AB}\frac{\partial^2 c_A}{\partial z^2} \text{ represents axial diffusion}$$

The relative magnitude of the two transport mechanisms in the axial direction can be compared by performing an order of magnitude analysis. Let L be a characteristic length in the z direction over which significant concentration changes occur. We can state

$$v_z \frac{\partial c_A}{\partial z} = O\left(<v_z> \frac{c_{As}}{L}\right) \tag{6.108}$$

$$D_{AB}\frac{\partial^2 c_A}{\partial z^2} = O\left(D_{AB}\frac{c_{As}}{L^2}\right) \quad (6.109)$$

We will explore the conditions for which convection dominates over conduction in the axial direction. According to the previous order of magnitude estimates, this will happen whenever

$$<v_z>\frac{c_{As}}{L} \gg D_{AB}\frac{c_{As}}{L^2} \quad (6.110)$$

Rearranging yields

$$Pe\frac{L}{2R} \gg 1 \quad (6.111)$$

where the Péclet number is defined by

$$Pe = \frac{<v_z>\, 2R}{D_{AB}} \quad (6.112)$$

Under these conditions, Equation 6.104 simplifies to

$$2<v_z>\left(1-\frac{r^2}{R^2}\right)\frac{\partial c_A}{\partial z} = D_{AB}\frac{1}{r}\frac{\partial}{\partial r}\left(r\frac{\partial c_A}{\partial r}\right) \quad (6.113)$$

where we have substituted the velocity field from Equation 6.102. This equation can be solved subject to boundary conditions 6.105 through 6.107.

Note that we would expect $L \gg 2R$ and thus moderate values of Pe might be enough to satisfy constraint (6.110). For example, if the solvent is water at 20°C flowing through a pipe with $R = 1$ cm at $<v_z> = 1$ cm/s, taking $D_{AB} = 10^{-5}$ cm^2/s (see Table 3.2 for liquid diffusivities) we would have $Pe = 2\times10^5$, which satisfies constraint (6.111). Under these conditions, $Re = 200$, so that the flow is laminar.

Next, we make the problem dimensionless by defining the following dimensionless variables:

$$\Theta = \frac{c_A}{c_{As}} \quad (6.114)$$

$$\eta = \frac{r}{R} \quad (6.115)$$

$$\xi = \frac{z}{PeR} \quad (6.116)$$

Using these definitions in Equation 6.113 leads to

$$(1-\eta^2)\frac{\partial \Theta}{\partial \xi} = \frac{1}{\eta}\frac{\partial}{\partial \eta}\left(\eta \frac{\partial \Theta}{\partial \eta}\right) \tag{6.117}$$

and the boundary conditions are

$$\Theta = 0, \xi = 0 \tag{6.118}$$

$$\frac{\partial \Theta}{\partial \eta} = 0, \eta = 0 \tag{6.119}$$

$$\Theta = 1, \eta = 1 \tag{6.120}$$

This problem does not have a known analytical solution. Hence, a solution to this problem should be found numerically. Here, we will find an analytical solution to a simplified version of the problem.

First, consider the expected shape for the radial concentration profiles along the pipe for the case (refer to Figure 6.19). The profile starts uniform at $z = 0$ (1), but the concentration at $r = R$ increases suddenly to c_{As}. At short distances from the entrance (2), the fluid at the core still will be at $c_A = 0$, but sharp concentration gradients occur near the wall. Further downstream (3), the concentration starts to increase even at the center of the pipe. At long distances from the entrance (4), the concentration will tend to approach the solubility everywhere.

From a mass transfer rate standpoint, the short-z region (region 2 in Figure 6.19, also called the entrance region) is the most important, since steep concentration profiles imply a faster mass transfer rate. In what follows, we will find a solution of the problem in this region. We will consider that the concentration reaches $c_A = 0$ at a relatively short distance from the wall. Later, we will establish quantitatively when this solution is applicable.

Since our analysis will be limited to a region close to the wall, we will use a new dimensionless coordinate that represents distances measured from the wall into the pipe. Let

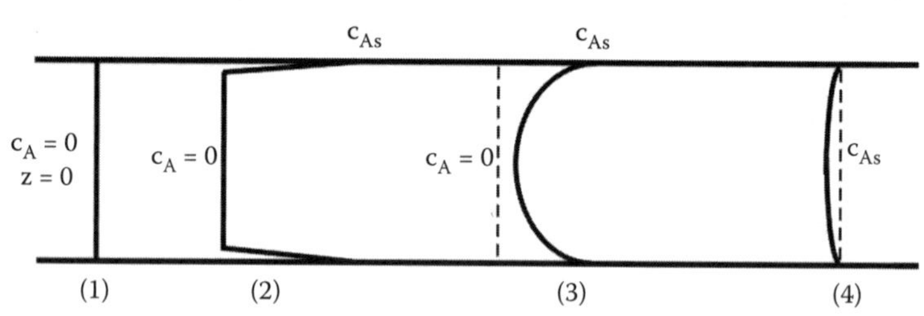

FIGURE 6.19 Evolution of radial concentration profiles along a pipe (see text for explanation).

$$\sigma = 1 - \eta \tag{6.121}$$

Our solution will be restricted to a region close to the wall: $\sigma \ll 1$. With this in mind, consider the approximations

$$\eta = 1 - \sigma \approx 1 \tag{6.122}$$

$$1 - \eta^2 = 1 - (1 - \sigma)^2 = 2\sigma - \sigma^2 \approx 2\sigma \tag{6.123}$$

Using the change of variables (6.121) and these approximations, the PDE (6.117) can be written as

$$2\sigma \frac{\partial \Theta}{\partial \xi} = \frac{\partial^2 \Theta}{\partial \sigma^2} \tag{6.124}$$

The boundary conditions (6.118) and (6.120) are rewritten as

$$\Theta = 0, \xi = 0 \tag{6.125}$$

$$\Theta = 1, \sigma = 0 \tag{6.126}$$

The boundary condition (6.119) will be out of the new domain. However, we know that, as we increase σ from the wall, the concentration will be eventually equal to $c_A = 0$ (which implies $\Theta = 0$). Hence, we can replace boundary condition (6.119) by

$$\Theta = 0, \sigma \to \infty \tag{6.127}$$

This problem can be solved analytically using the method of **combination of variables**, also called **similarity transformation**. The technique is based on the postulate that there is an independent variable s (called the **similarity variable**) that is a combination of ξ and σ such that $\Theta = \Theta(s)$. The mathematical form proposed for the similarity variable is

$$s = \sigma \xi^n \tag{6.128}$$

To test if this will work, we transform the differential equation, for which we need the following derivatives, obtained using the chain rule,

$$\frac{\partial \Theta}{\partial \xi} = \frac{d\Theta}{ds}\frac{\partial s}{\partial \xi} = n\sigma\xi^{n-1}\frac{d\Theta}{ds} \tag{6.129}$$

$$\frac{\partial \Theta}{\partial \sigma} = \frac{d\Theta}{ds}\frac{\partial s}{\partial \sigma} = \xi^n \frac{d\Theta}{ds} \tag{6.130}$$

$$\frac{\partial^2\Theta}{\partial\sigma^2} = \xi^n \frac{\partial}{\partial\sigma}\left(\frac{d\Theta}{ds}\right) = \xi^n \frac{d^2\Theta}{ds^2}\frac{\partial s}{\partial\sigma} = \xi^{2n}\frac{d^2\Theta}{ds^2} \tag{6.131}$$

Substituting Equations 6.129 and 6.131 into Equation 6.124 leads to

$$2n\sigma^2\xi^{-n-1}\frac{d\Theta}{ds} = \frac{d^2\Theta}{ds^2} \tag{6.132}$$

Now we note that, for Θ to be an exclusive function of s, the variables σ and ξ must disappear from the differential equation. We substitute σ in terms of s from Equation 6.128 to get

$$2ns^2\xi^{-3n-1}\frac{d\Theta}{ds} = \frac{d^2\Theta}{ds^2} \tag{6.133}$$

Now, we let $n = -1/3$ to eliminate ξ and obtain

$$-\frac{2}{3}s^2\frac{d\Theta}{ds} = \frac{d^2\Theta}{ds^2} \tag{6.134}$$

We still need to check if our similarity variable

$$s = \frac{\sigma}{\xi^{1/3}} \tag{6.135}$$

will lead to a consistent formulation of the boundary conditions. First, we notice that conditions 6.125 and 6.127 both lead to

$$\Theta = 0, s \to \infty \tag{6.136}$$

whereas condition 6.126 leads to

$$\Theta = 1, s = 0 \tag{6.137}$$

This problem satisfies the postulate $\Theta = \Theta(s)$, since we have eliminated all reference to the original independent variables. Now, we need to solve the ODE 6.134 subject to the boundary conditions 6.136 and 6.137. To integrate the ODE, let

$$u = \frac{d\Theta}{ds} \Rightarrow \frac{du}{ds} = -\frac{2}{3}s^2u \tag{6.138}$$

Separating and integrating leads to

$$\ln u = -\frac{2}{9}s^3 + A \Rightarrow u = C\exp\left(-\frac{2s^3}{9}\right) = \frac{d\Theta}{ds} \tag{6.139}$$

Separating and integrating again, using boundary condition 6.137 to make the integral definite leads to

$$\int_1^{\Theta} d\Theta = C\int_0^{s} \exp\left(-\frac{2s^3}{9}\right)ds \tag{6.140}$$

$$\Rightarrow \Theta = 1 + C\int_0^{s} \exp\left(-\frac{2s^3}{9}\right)ds \tag{6.141}$$

Applying boundary condition 6.136 yields

$$0 = 1 + C\int_0^{\infty} \exp\left(-\frac{2s^3}{9}\right)ds \Rightarrow C = -\frac{1}{\int_0^{\infty} \exp\left(-2s^3/9\right)ds} \tag{6.142}$$

and the solution becomes

$$\Theta = 1 - \frac{\int_0^{s} \exp\left(-2s^3/9\right)ds}{\int_0^{\infty} \exp\left(-2s^3/9\right)ds} \tag{6.143}$$

The integrals in the solution do not have a known analytical form, but they can be evaluated numerically. Evaluating the integral in the denominator leads to

$$\Theta = 1 - 0.6783\int_0^{s} \exp\left(-\frac{2s^3}{9}\right)ds \tag{6.144}$$

This solution is plotted in Figure 6.20. Note that this solution is valid only for $\sigma \ll 1$. In view of the functional form of the similarity variable (Equation 6.135), and observing that the relevant domain in s extends to $s \approx 3$ (Figure 6.20), we conclude that ξ must be small also. However, recall that ξ is inversely proportional to Pe, and that we are seeking a solution valid for large Pe. This means that the solution given by Equation 6.144 should extend at least to an axial distance equal to several pipe diameters if Pe is large enough. We will explore this in more detail below.

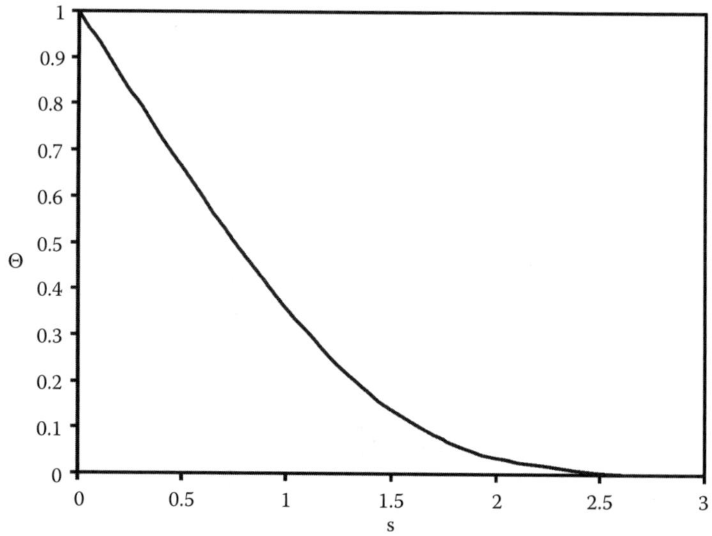

FIGURE 6.20 Similarity solution for the dimensionless temperature profile in the entrance region of a pipe with a sudden change in wall temperature.

The dissolution flux is given by

$$-N_{Ar}\Big|_{r=R} = D_{AB}\frac{\partial c_A}{\partial r}\Big|_{r=R} \tag{6.145}$$

Using the solution 6.144 we obtain, after manipulations,

$$-N_{Ar}\Big|_{r=R} = 0.6783\frac{D_{AB}}{R}c_{As}\left(\frac{RPe}{z}\right)^{1/3} \tag{6.146}$$

Recalling that $c_A = 0$ away at the core of the pipe, we define now a convective mass transfer coefficient as

$$-N_{Ar}\Big|_{r=R} = k_c c_{As} \tag{6.147}$$

which expresses the species molar flux as proportional to the driving force for convective mass transfer, which is the concentration difference between the wall region and the bulk (pipe core) region. From the last two equations we find,

$$Sh = \frac{k_c(2R)}{D_{AB}} = 1.357\left(\frac{RPe}{z}\right)^{1/3} \tag{6.148}$$

The left-hand side of this equation is the Sherwood number. The Péclet number is usually expressed in terms of the Reynolds number

$$\mathrm{Re} = \frac{\rho <v_z> 2R}{\mu} \tag{6.149}$$

and the Schmidt number, defined as

$$\mathrm{Sc} = \frac{\mu}{\rho D_{AB}} \tag{6.150}$$

Note that

$$\mathrm{Pe} = \mathrm{ReSc} \tag{6.151}$$

so that Equation 6.148 is usually written as

$$\mathrm{Sh} = 1.077 \left(\frac{2R}{z} \right)^{1/3} (\mathrm{ReSc})^{1/3} \tag{6.152}$$

Next, we assess the conditions under which the previous analysis is valid. For this purpose, we would like to define the region adjacent to the wall in which the concentration changes appreciably. In terms of our solution, Figure 6.20, this region is $0 < s < \infty$, since Θ reaches zero only asymptotically (Figure 6.20). However, we can set a practical limit to define the region of change. For example, we can state that, when $\Theta = 0.01$, the liquid is free of solute for all practical purposes. For Equation 5.144, we can calculate numerically that $\Theta = 0.01$ at $s = 2.31$. Combining Equations 6.116 and 6.135, we see that this corresponds to a value of σ given by

$$\sigma = 2.31 \left(\frac{z}{\mathrm{PeR}} \right)^{1/3} \tag{6.153}$$

Since the entrance region requires that $\sigma \ll 1$, this means that the solution will be valid if

$$2.31 \left(\frac{z}{\mathrm{PeR}} \right)^{1/3} \ll 1, \text{entrance region} \tag{6.154}$$

Or

$$\frac{z}{R} \ll 0.0811\ \mathrm{Pe}, \text{entrance region} \tag{6.155}$$

Notice that, since our solution is applicable for relatively large values of the Peclet number (Equation 6.111), this would give an entrance region that is several times the pipe radius. For example, if the solvent is water at 20°C flowing through a pipe with R = 1 cm, we calculated before that $Pe = 2 \times 10^5$, then $z \ll 1.62 \times 10^4 R$ defines the entrance region. This ensures that the solution is valid in this case for z values that can be up to perhaps a hundred times the radius.

6.5 CONVECTIVE MASS TRANSFER CORRELATIONS

The previous example showed how the solution of the convective–diffusion equation can be used to predict the convective mass transfer coefficient. For complex flows, solution of the convective–diffusion equation must be done numerically and mass transfer rates are not calculated using an explicit relation. In these cases, calculations of mass transfer rates rely on the use of either empirical of fitted relations to calculate the mass transfer coefficient. These correlations must be considered valid only inside the range of experimental parameters for which they were developed. Examples of correlations for specific cases follow.

Turbulent flow in a pipe (Gilliland and Sherwood correlation):

$$Sh = 0.023\, Re^{0.83}\, Sc^{0.44} \tag{6.156}$$

Range of applicability: $2000 < Re < 35{,}000$, $0.6 < Sc < 2.5$.

Flow around a sphere

1. Frössling correlation:

$$Sh = 2 + 0.552\, Re^{0.5}\, Sc^{0.33} \tag{6.157}$$

   Range of applicability: $Re < 1$, $0.6 < Sc < 3200$.
2. Treybal's correlation:

$$Sh = 2 + 0.347(Re\, Sc^{1/2})^{0.62} \tag{6.158}$$

   Range of applicability: $Re < 3 \times 10^4$, $0.6 < Sc < 3200$. Note that, as $Re \to 0$, this equation satisfies what is expected at the diffusive limit (Equation 4.63).
3. Inertial flow:

$$Sh = 0.347\, Re^{0.62}\, Sc^{0.31} \tag{6.159}$$

   Range of applicability: $2000 < Re < 3 \times 10^4$, $0.6 < Sc < 3200$.

Transverse flow around a cylinder

$$Sh = (0.4\, Re^{1/2} + 0.06\, Re^{2/3})Sc^{0.4} \tag{6.160}$$

Range of applicability: $Re < 10^5$.

EXAMPLE 6.1 DISSOLUTION OF A SPHERICAL PARTICLE IN A MOVING FLUID

In Section 4.2.2 we modeled the dissolution of a solid sphere in a stagnant liquid. In that case, mass transfer in the liquid phase was by diffusion only. Here, we will extend the analysis to a fluid flowing with a Reynolds and Schmidt numbers within the range of applicability of Treybal's correlation (Equation 6.158). The dissolution flux from the surface of the sphere is represented by the convective mass transfer coefficient,

$$N_{Ar}\big|_{r=R} = k_c(c_{As} - c_{A\infty}) \tag{6.161}$$

where c_{As} is the solubility of the solid (A) in the liquid (B) and $c_{A\infty}$ is the concentration of the solute far from the sphere. In this case, we will take $c_{A\infty} = 0$.

The mass transfer coefficient will be predicted from Treybal's correlation (Equation 6.158) which can be written as follows, after substitution of the definitions of the dimensionless groups,

$$\frac{2Rk_c}{D_{AB}} = 2 + 0.347\left(\frac{2R\rho_l v_l}{\mu_l}\right)^{0.62}\left(\frac{\mu_l}{\rho_l D_{AB}}\right)^{0.31} \tag{6.162}$$

where v_l is the liquid velocity far from the sphere, ρ_l and μ_l are the liquid density and viscosity, respectively (note that these are properties of the liquid phase, since it is in the liquid where convective mass transfer is taking place), and D_{AB} is the diffusivity of the solute in the liquid phase. Substitution of Equation 6.162 into Equation 6.161 leads to, after manipulations,

$$k_c = \frac{D_{AB}}{R} + 0.267\left(\frac{\rho_l}{\mu_l}\right)^{0.31} v_l^{0.62} D_{AB}^{0.69} R^{-0.38} \tag{6.163}$$

and the molar dissolution flux (Equation 6.162) is

$$N_{Ar}\big|_{r=R} = \left[\frac{D_{AB}}{R} + 0.267\left(\frac{\rho_l}{\mu_l}\right)^{0.31} v_l^{0.62} D_{AB}^{0.69} R^{-0.38}\right] c_{As} \tag{6.164}$$

This equation can be used to predict how the radius of the sphere decreases with time. For that purpose, we use a mole balance taking the solid sphere as control volume

$$\frac{d(c_s V)}{dt} = -N_{Ar}\big|_{r=R} 4\pi R^2 \tag{6.165}$$

where c_s is the molar density of the solid. Using $V = 4\pi R^3/3$, this equation can be written as

$$\frac{dR}{dt} = -\frac{1}{c_s} N_{Ar}\big|_{r=R} \tag{6.166}$$

The molar dissolution flux is substituted from Equation 6.164, which leads to

$$\frac{dR}{dt} = -\frac{\alpha}{R} - \beta R^{-0.38} \tag{6.167}$$

where the constants α and β are given by

$$\alpha = \frac{D_{AB} c_{As}}{c_s} \tag{6.168}$$

$$\beta = 0.267 \left(\frac{\rho_l}{\mu_l} \right)^{0.31} v_l^{0.62} D_{AB}^{0.69} \frac{c_{As}}{c_s} \tag{6.169}$$

Equation 6.167 can be separated and integrated from $t = 0$, $R = R_0$ to an arbitrary time t

$$\int_{R_0}^{R} \frac{dR}{(\alpha/R) + \beta R^{-0.38}} = -t \tag{6.170}$$

which gives R(t). The integral in this equation must be found numerically.

For example, consider the dissolution of a sphere of benzoic acid in pure water at 20°C. The sphere has initially a radius $R_0 = 0.5$ cm. We will calculate how the radius changes with time when the dissolution occurs in a stagnant liquid and in a liquid moving with a velocity $v_l = 5$ cm/s. At 20°C, the solubility of benzoic acid in water is $c_{As} = 2.7$ mol/m³. Other relevant properties are: $\rho_l = 1000$ kg/m³, $\mu_l = 1.0$ mPa s, $D_{AB} = 0.9 \times 10^{-9}$ m²/s, $\rho_s = 1270$ kg/m³ (density of solid benzoic acid), $M_A = 122.12$ g/mol. The molar density of the solid is

$$c_s = \frac{\rho_s}{M_A} = 10.4 \frac{\text{kmol}}{\text{m}^3} \tag{6.171}$$

The Schmidt number and the Reynolds number at the initial radius can be calculated from Equations 6.149 and 6.150 to obtain: Sc = 1111, Re = 500, which ensures that we are within the applicable range of Treybal's correlation. Now the parameters α and β can be calculated and Equation 6.170 can be used to find R(t).

Figure 6.21 shows the results of the calculation. It takes about 300 h to dissolve the sphere completely. In this time scale, the radius of the sphere would not decrease appreciably from the initial value if the liquid were stagnant (as predicted by Equation 4.52).

PROBLEMS

6.1 A large solid block (see figure) slides down a slope on the surface of a liquid (density ρ and viscosity μ) at a constant velocity v_0. Because upstream and downstream from the block the liquid film is exposed to the atmosphere, you may assume that pressure in the liquid between the block and

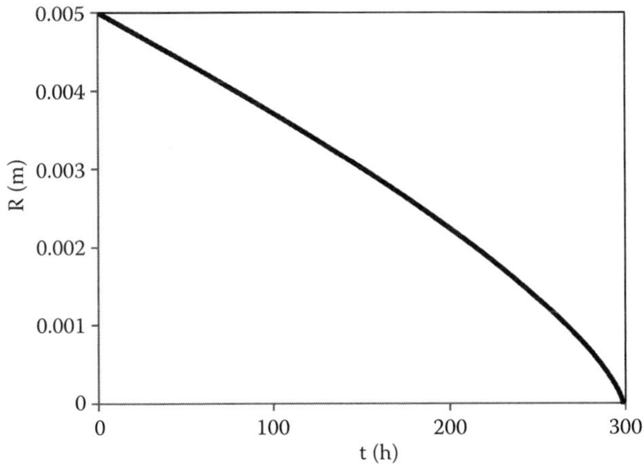

FIGURE 6.21 Dissolution of a benzoic acid sphere in moving water.

the solid surface does not change with x. You also may assume that the liquid film has a uniform thickness δ.

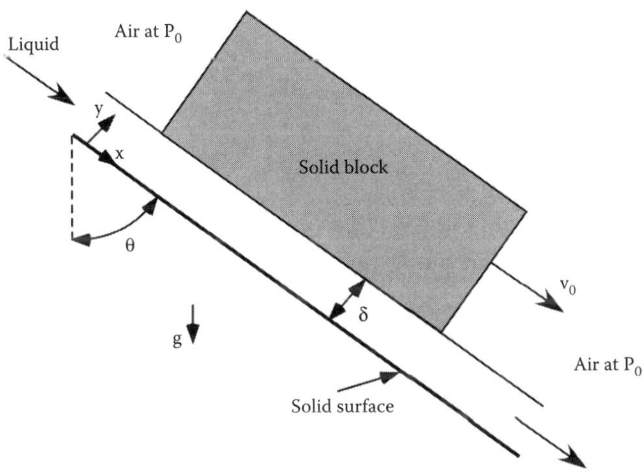

a. Determine the velocity profile $v_x(y)$ in the liquid contained between the block and the solid surface. Assume that the flow is steady and incompressible. Sketch the profile.
b. If the surface area of the block that is in contact with the liquid is A_0, find the force that the liquid exerts on the block in the x direction.
c. If the mass of the block is M, find an expression to calculate v_0.

6.2 A solid plate slides down an inclined plane at a constant velocity v_0 on top of a uniform liquid film of thickness δ (see figure). The downward motion of the liquid is due to gravitational forces and the motion of the plate; that is, there is no pressure variation in the direction of motion (x). The liquid density and viscosity are ρ and μ, respectively.

a. Find the velocity field in the liquid film and the volumetric flow rate (consider a depth W in the z direction).
b. Find the value of v_0 that would make the volumetric flow rate in the film equal to that of a film with a free surface (i.e., when the fluid surface at $y = \delta$ is open to the atmosphere). What is the force per unit area that the fluid exerts on the moving plate in the x direction under this condition?

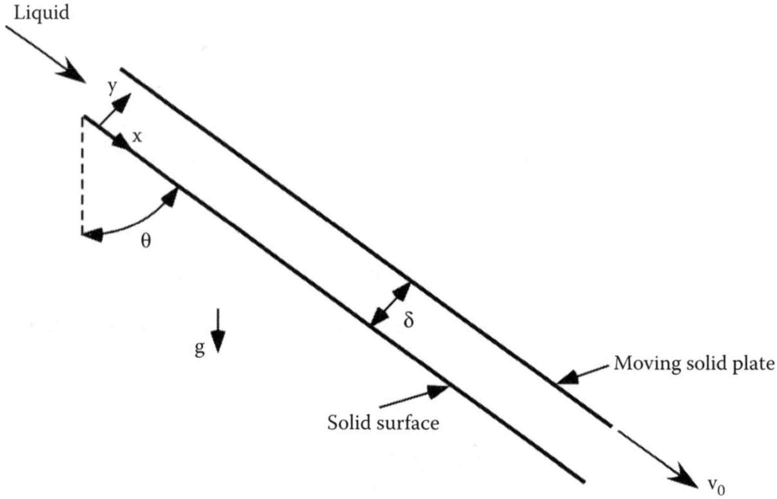

6.3 A thin water film (thickness δ_1) forms on the surface of a flat roof as shown in the figure. As the liquid approaches the edge of the roof, it falls down forming a thin film on the sidewall. If the thickness of the film on the roof is $\delta_1 = 3$ mm, and $\theta = 60°$, determine δ_2.

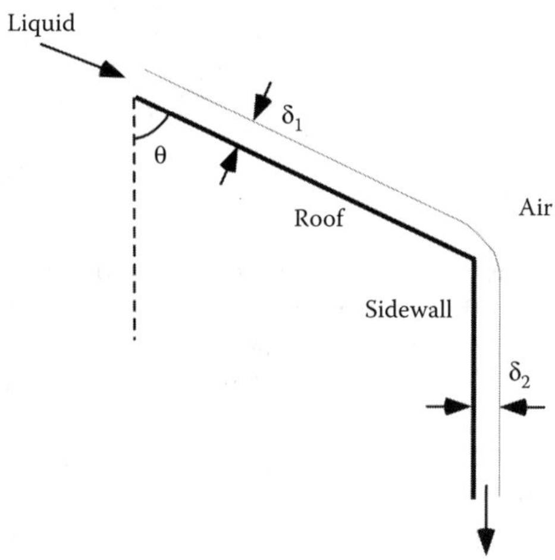

6.4 A liquid (density ρ and viscosity μ) is placed between two large vertical plates as shown in the figure. The plate on the left is stationary and the plate on the right moves up with a constant velocity v_0. The device is open to the atmosphere, so that the pressure in the liquid can be considered equal to atmospheric pressure (P_0) everywhere. Take the depth of the system (z direction) to be W.

a. Find the velocity profile, $v_x(y)$.
b. Find an expression to calculate the volumetric flow rate in the x direction, Q. Determine the value of v_0 that would make $Q = 0$. Sketch the velocity profile between the plates for this case.
c. Find an expression to calculate the vertical force per unit area necessary to keep the moving plate ($y = h$) at constant velocity v_0 (neglect the weight of the plate). Find also the force per unit area that the fluid exerts on the stationary plate ($y = 0$).

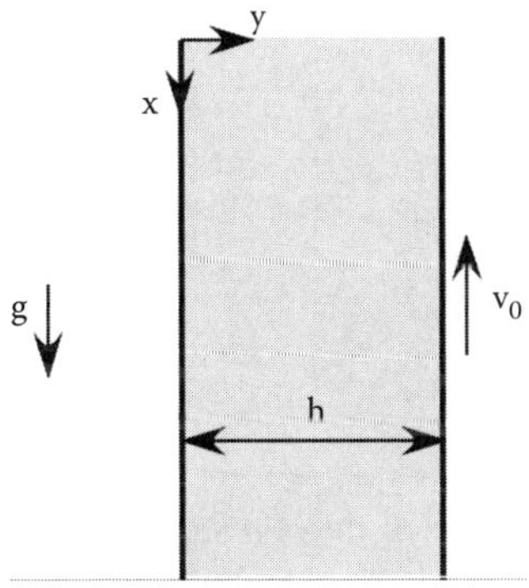

6.5 A cylindrical pipe ($R = 1$ cm) is used to transport a solution that is supersaturated with a solid. Small crystals (microns in size) of the solid eventually deposit on the internal surface of the pipe. Independent experiments have shown that a shear stress of 0.02 Pa is strong enough to detach the crystals from the solid surface. To clean the pipe, it is proposed to flow water at 20°C ($\mu = 10^{-3}$ Pa s) through it at a flow rate high enough to detach the crystals from the pipe wall and flush them away. Determine the volumetric flow rate of water necessary to clean the pipe.

6.6 A small cylindrical pipe with radius $R = 1$ cm and length $L = 10$ cm is completely filled with a viscous liquid ($\rho = 890$ kg/m^3, $\mu = 0.08$ Pa s) and closed at both ends. The pipe is held vertically and both ends ($z = 0$ and $z = L$) are suddenly opened to atmospheric pressure so that the liquid starts to flow out of the pipe through the lower end. Assuming that the flow in the pipe is incompressible and becomes steady instantaneously, estimate the time required to drain all the liquid from the pipe.

6.7 A wetted-wall column is a common absorber used in small-scale applications. It consists of a cylindrical pipe on whose inner surface the liquid forms a uniform film with thickness $R_2 - R_1$ (see figure). The liquid film flows downward due to gravity, and gas is made to flow countercurrently to the liquid. Under normal operating conditions, the gas velocity is so small that its pressure is uniform and equal to atmospheric pressure, P_a, and the gas does not exert an appreciable shear stress on the liquid surface. For a liquid of known density and viscosity (ρ, μ), find the velocity profile, $v_z(r)$.

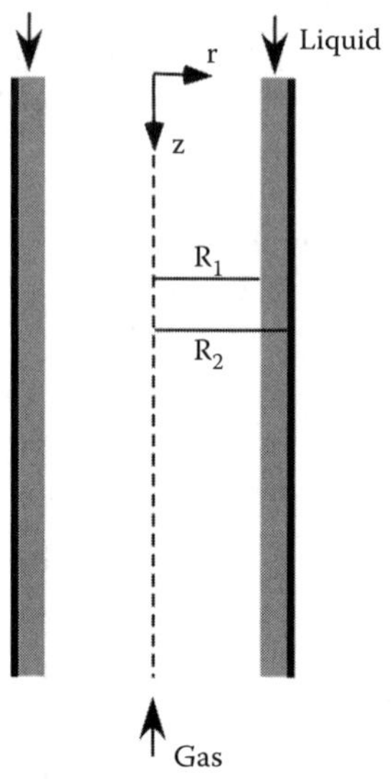

6.8 Pasta can be bought fresh or dry. Fresh pasta has a high moisture content, which makes it susceptible to grow microorganisms. Because of its long shelf life, dry pasta is widely available. Spaghetti strands coming out of an extrusion process are dried by exposing them to a continuous flow of relatively dry air. Drying would be faster at high temperatures but temperatures above 60°C are known to modify the starches in pasta to a point that flavor and texture are compromised.

Consider a long, cylindrical spaghetti strand (D = 2.4 mm). The spaghetti comes out of the extruder with a moisture content $c_{A0} = 0.55$ g water/cm^3. A flow of air at 55°C and relative humidity of 5% is established perpendicular to the spaghetti. The air velocity far from the spaghetti is 10 cm/s. For a fluid flowing perpendicular to a cylinder in laminar flow, the mass transfer coefficient can be estimated from the empirical correlation

$$Sh = (0.4\,Re^{1/2} + 0.06\,Re^{2/3})Sc^{0.4}$$

where the velocity in the Reynolds number is the velocity far from the cylinder, and the characteristic length used in the definition of the Reynolds and Sherwood numbers is the cylinder diameter.

Estimate the time that the spaghetti must spend in the air flow to remove 80% of its initial water content.

Physical properties:

Atmospheric pressure: 1 atm
Diffusivity of water vapor in air at 55°C: 0.30 cm^2/s
Air density at 55°C: 1.0 kg/m^3
Air viscosity at 55°C: 2.0×10^{-5} Pa s
Diffusivity of liquid water in spaghetti: 2.5×10^{-7} cm^2/s
Vapor pressure of water at 55°C: 0.155 atm

The partial pressure of water in air that is at equilibrium with spaghetti containing a water concentration c_A is given by $P_A = Hc_A$, with $H = 0.3$ atm cm^3/g at 55°C.

7 Hydrodynamic Dispersion

7.1 THE CONCEPT OF HYDRODYNAMIC DISPERSION AND TAYLOR–ARIS THEORY

In this section, we consider the combined effects of diffusion and convection on the transport of a solute in a nonuniform flow. The solute (A) is assumed to be under dilute conditions, so that the convection–diffusion equation describes the concentration field. As expected, the distribution of solute will be affected by the velocity field. If the flow is one-dimensional and uniform (Figure 7.1a), a slug of solute will move uniformly with the fluid and eventually, diffusion will cause it to spread. If the flow is one-dimensional, but nonuniform (Figure 7.1b), a slug of solute will spread and change its shape due to the existence of velocity gradients.

We are interested in the transport of a solute in a pipe under unsteady conditions. If the flow in the pipe is laminar and developed, the parabolic velocity profile will affect the solute distribution in the system, in a manner analogous to the example shown in Figure 7.1b. Note that, even if the concentration of solute is uniform over the cross section at a specific axial position in the pipe, we must consider that $c_A = c_A(t,r,z)$ in general, due to the effects of the velocity field on transport (angular symmetry might be assumed if it is not broken by boundary conditions). For developed flow, the convection–diffusion equation in cylindrical coordinates simplifies to

$$\frac{\partial c_A}{\partial t} + v_z \frac{\partial c_A}{\partial z} = D_A \left[\frac{1}{r} \frac{\partial}{\partial r} \left(r \frac{\partial c_A}{\partial r} \right) + \frac{\partial^2 c_A}{\partial z^2} \right] + R_A \tag{7.1}$$

where

$$v_z = 2 \langle v_z \rangle \left[1 - \left(\frac{r}{R} \right)^2 \right] \tag{7.2}$$

Equation 7.1 would have to be solved with appropriate boundary conditions to find the concentration field. Consider now that we are interested in applications that involve the transport of solute over long stretches of pipe (i.e., over lengths of pipe that are much larger than the pipe diameter). Variations of c_A with radial position

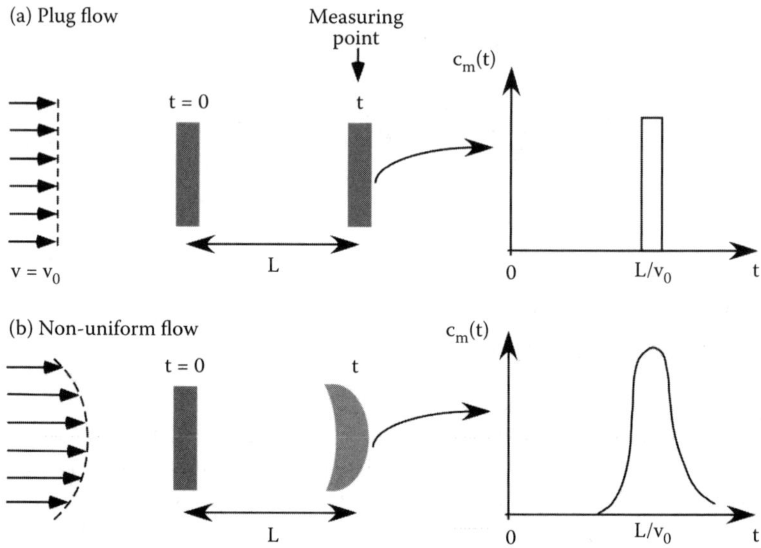

FIGURE 7.1 A slug of solute is released at t = 0 at a point in a moving fluid. At a distance L downstream from the release point, the average concentration of solute over the flow cross section is measured as a function of time, $c_m(t)$. If the velocity is uniform over the cross section, the measured concentration would be a perfect square function (a). If the velocity field is nonuniform (as in b), there will be a spread of solute in the direction of motion. The measured concentration in the latter case will have the shape shown. Here, we are neglecting the spreading of the solute due to diffusion.

will not be of interest in that case. The use of a cross-sectional average concentration would be more appropriate,

$$<c_A> = \frac{1}{A}\int_A c_A dA \tag{7.3}$$

where A is the pipe's cross-sectional area. Due to the angular symmetry, we can use $dA = 2\pi r\, dr$, with $A = \pi r^2$, so that Equation 7.3 can be written as

$$<c_A> = \frac{2}{R^2}\int_0^R c_A r dr \tag{7.4}$$

Note that $<c_A> = <c_A>(t,z)$. To find a governing equation for $<c_A>$, we take the cross-sectional average of Equation 7.1:

$$\frac{2}{R^2}\int_0^R \left[\frac{\partial c_A}{\partial t} + v_z \frac{\partial c_A}{\partial z}\right] r dr = \frac{2}{R^2}\int_0^R D_A \left[\frac{1}{r}\frac{\partial}{\partial r}\left(r\frac{\partial c_A}{\partial r}\right) + \frac{\partial^2 c_A}{\partial z^2}\right] r dr + \frac{2}{R^2}\int_0^R R_A r dr \tag{7.5}$$

Term by term, we have

$$\frac{2}{R^2}\int_0^R \frac{\partial c_A}{\partial t} r dr = \frac{\partial}{\partial t}\left(\frac{2}{R^2}\int_0^R c_A r dr\right) = \frac{\partial <c_A>}{\partial t} \tag{7.6}$$

$$\frac{2}{R^2}\int_0^R v_z \frac{\partial c_A}{\partial z} r dr = \frac{\partial}{\partial z}\left[\frac{2}{R^2}\int_0^R v_z c_A r dr\right] = \frac{\partial <v_z c_A>}{\partial z} \tag{7.7}$$

(note that in this equation we have used the fact that v_z is independent of z). Also, if D_A is uniform, we have

$$\frac{2}{R^2}\int_0^R D_A\left[\frac{1}{r}\frac{\partial}{\partial r}\left(r\frac{\partial c_A}{\partial r}\right)\right] r dr = \frac{2D_A}{R^2}\int_0^R \frac{\partial}{\partial r}\left(r\frac{\partial c_A}{\partial r}\right) dr = \frac{2D_A}{R^2}\left(r\frac{\partial c_A}{\partial r}\right)_{r=0}^{r=R} \tag{7.8}$$

Due to angular symmetry, we can state that

$$\frac{\partial c_A}{\partial r} = 0,\ r = 0 \tag{7.9}$$

In addition, if the solute is not transported into the wall of the pipe, its molar flux at r = R is zero, which implies

$$\frac{\partial c_A}{\partial r} = 0,\ r = R \tag{7.10}$$

Hence,

$$\frac{2}{R^2}\int_0^R D_A\left[\frac{1}{r}\frac{\partial}{\partial r}\left(r\frac{\partial c_A}{\partial r}\right)\right] r dr = 0 \tag{7.11}$$

Also,

$$\frac{2}{R^2}\int_0^R D_A \frac{\partial^2 c_A}{\partial z^2} r dr = D_A \frac{\partial^2}{\partial z^2}\left(\frac{2}{R^2}\int_0^R c_A r dr\right) = D_A \frac{\partial^2 <c_A>}{\partial z^2} \tag{7.12}$$

Finally, the last term in Equations 7.5 is

$$\frac{2}{R^2}\int_0^R R_A r dr = <R_A> \tag{7.13}$$

Collecting terms, the averaged convective diffusion equation is

$$\frac{\partial <c_A>}{\partial t} + \frac{\partial <v_z c_A>}{\partial z} = D_A \frac{\partial^2 <c_A>}{\partial z^2} + <R_A> \tag{7.14}$$

At this point, this equation cannot be used to find the average concentration, owing to the presence of the term $<v_z c_A>$. We will rewrite this equation by using the concentration and velocity deviations, defined by

$$\tilde{c}_A = c_A - <c_A> \tag{7.15}$$

$$\tilde{v}_z = v_z - <v_z> \tag{7.16}$$

From these two equations, we can see that

$$v_z c_A = <c_A><v_z> + \tilde{c}_A <v_z> + <c_A> \tilde{v}_z + \tilde{v}_z \tilde{c}_A \tag{7.17}$$

Taking the average of this equation and noticing that average concentrations are constants in the averaging process and that the average of a deviation variable is zero (which can be shown by averaging Equations 7.15 and 7.16), we get

$$<v_z c_A> = <c_A><v_z> + <\tilde{v}_z \tilde{c}_A> \tag{7.18}$$

Substituting this equation into Equation 7.14 leads to

$$\frac{\partial <c_A>}{\partial t} + <v_z> \frac{\partial <c_A>}{\partial z} = D_A \frac{\partial^2 <c_A>}{\partial z^2} + <R_A> - \frac{\partial <\tilde{v}_z \tilde{c}_A>}{\partial z} \tag{7.19}$$

The last term of this equation now prevents us from attempting a solution having $<c_A>$ and a dependent variable. In addition, the reaction term will be a problem if the kinetic rate expression is nonlinear. For a first-order consumption reaction, this will not be a problem since

$$<R_A> = -k<c_A> \tag{7.20}$$

In what follows, for simplicity, we will consider the case of **passive dispersion**, in which the solute does not undergo chemical transformations and $R_A = 0$.

Consider the analysis of transport of solute over a pipe length scale L such that

$$L \gg R \tag{7.21}$$

using the average concentration as a measure of solute content. A mathematical comparison between Equation 7.19 and the original convection–diffusion equation differs by the last term on the right-hand side. Since Equation 7.19 represents a mole balance of species A, that term can be conceptualized as an **apparent** mass transport mechanism that is due to the existence of deviations of velocity and concentration. This mechanism is called **dispersion**. When applied to flow in pipes, it is usually termed **Taylor dispersion**. To find closure to this problem, we will seek a representation for the concentration deviation. This will be possible only if certain simplifying assumptions are made. To start the analysis, we subtract Equation 7.19 from the point Equation 7.1. Using Equation 7.15, this leads to, after simplifications (recall that we are considering now the case in which there is no reaction),

$$\frac{\partial \tilde{c}_A}{\partial t} + v_z \frac{\partial c_A}{\partial z} - \langle v_z \rangle \frac{\partial \langle c_A \rangle}{\partial z} = D_A \left[\frac{1}{r} \frac{\partial}{\partial r} \left(r \frac{\partial \tilde{c}_A}{\partial r} \right) + \frac{\partial^2 \tilde{c}_A}{\partial z^2} \right] + \frac{\partial \langle \tilde{v}_z \tilde{c}_A \rangle}{\partial z} \tag{7.22}$$

Using Equations 7.15 and 7.16, we have

$$v_z \frac{\partial c_A}{\partial z} = \langle v_z \rangle \frac{\partial \langle c_A \rangle}{\partial z} + \tilde{v}_z \frac{\partial \langle c_A \rangle}{\partial z} + v_z \frac{\partial \tilde{c}_A}{\partial z} \tag{7.23}$$

and Equation 7.22 becomes

$$\frac{\partial \tilde{c}_A}{\partial t} + \tilde{v}_z \frac{\partial \langle c_A \rangle}{\partial z} + v_z \frac{\partial \tilde{c}_A}{\partial z} - \frac{\partial \langle \tilde{v}_z \tilde{c}_A \rangle}{\partial z} = D_A \left[\frac{1}{r} \frac{\partial}{\partial r} \left(r \frac{\partial \tilde{c}_A}{\partial r} \right) + \frac{\partial^2 \tilde{c}_A}{\partial z^2} \right] \tag{7.24}$$

To simplify this equation, we perform an order of magnitude analysis. First, consider the two diffusion terms in this equation,

$$\frac{D_A}{r} \frac{\partial}{\partial r} \left(r \frac{\partial \tilde{c}_A}{\partial r} \right) = O\left(\frac{D_A \tilde{c}_A}{R^2} \right) \tag{7.25}$$

$$D_A \frac{\partial^2 \tilde{c}_A}{\partial z^2} = O\left(\frac{D_A \tilde{c}_A}{L^2} \right) \tag{7.26}$$

Clearly, if constraint 7.21 is satisfied, the axial diffusion term is negligible and Equation 7.24 simplifies to

$$\frac{\partial \tilde{c}_A}{\partial t} + \tilde{v}_z \frac{\partial \langle c_A \rangle}{\partial z} + v_z \frac{\partial \tilde{c}_A}{\partial z} - \frac{\partial \langle \tilde{v}_z \tilde{c}_A \rangle}{\partial z} = D_A \frac{1}{r} \frac{\partial}{\partial r} \left(r \frac{\partial \tilde{c}_A}{\partial r} \right) \tag{7.27}$$

Now consider the transient term; its order of magnitude is

$$\frac{\partial \tilde{c}_A}{\partial t} = O\left(\frac{\tilde{c}_A}{t_s} \right) \tag{7.28}$$

where t_s is a characteristic time over which significant changes in concentration occur. Comparison of Equations 7.25 and 7.28 indicates that the transient term would be negligible if

$$\frac{D_A t_s}{R^2} \gg 1 \tag{7.29}$$

Taking this constraint to be valid, Equation 7.27 simplifies to

$$\tilde{v}_z \frac{\partial <c_A>}{\partial z} + v_z \frac{\partial \tilde{c}_A}{\partial z} - \frac{\partial <\tilde{v}_z \tilde{c}_A>}{\partial z} = D_A \frac{1}{r}\frac{\partial}{\partial r}\left(r \frac{\partial \tilde{c}_A}{\partial r} \right) \tag{7.30}$$

The deviation velocity can be found from Equations 7.2 and 7.16, which yields

$$\tilde{v}_z = <v_z> \left[1 - 2\left(\frac{r}{R} \right)^2 \right] \tag{7.31}$$

It should be clear that both the point and deviation velocities are of the order of the average velocity,

$$\tilde{v}_z, v_z = O(<v_z>) \tag{7.32}$$

so that the second and third terms on the left-hand side of Equation 7.30 are of the order

$$\frac{\partial <\tilde{v}_z \tilde{c}_A>}{\partial z}, v_z \frac{\partial \tilde{c}_A}{\partial z} = O\left(\frac{<v_z>\tilde{c}_A}{L} \right) \tag{7.33}$$

Comparing Equations 7.25 and 7.33, we can see that these terms are negligible in Equation 7.30 whenever

$$\frac{D_A L}{R^2 <v_z>} \gg 1 \tag{7.34}$$

This constraint is usually expressed in terms of the Peclet number, defined by

$$Pe = \frac{<v_z>R}{D_A} \tag{7.35}$$

so that constraint (7.34) is

$$Pe\left(\frac{R}{L} \right) \ll 1 \tag{7.36}$$

Note that this constrain will be satisfied even for moderate Peclet numbers by virtue of constraint 5.193. Equation 5.202 now becomes

$$\tilde{v}_z \frac{\partial <c_A>}{\partial z} = D_A \frac{1}{r}\frac{\partial}{\partial r}\left(r \frac{\partial \tilde{c}_A}{\partial r} \right) \tag{7.37}$$

Since the derivative of the average concentration with respect to z is independent of r, Equation 7.37 suggests a solution for the deviation concentration given by

$$\tilde{c}_A = \frac{\partial <c_A>}{\partial z} f(r) \tag{7.38}$$

Substitution of Equations 7.31 and 7.38 into Equation 7.37 yields, after simplification,

$$\frac{1}{r}\frac{d}{dr}\left(r \frac{df}{dr} \right) = \frac{<v_z>}{D_A}\left[1 - 2\left(\frac{r}{R} \right)^2 \right] \tag{7.39}$$

Integrating this equation twice leads to

$$f = \alpha + \beta \ln r + \frac{<v_z>}{4D_A}\left(r^2 - \frac{r^4}{2R^2} \right) \tag{7.40}$$

where α and β are integration constants. Since the concentration must be finite at $r = 0$, this necessarily implies that $\beta = 0$. A second condition for f can be found by recognizing that the average of Equation 7.15 implies that **z** and, therefore,

$$<f> = 0 \tag{7.41}$$

Finding α by applying this condition to Equation 7.40 yields

$$f = \frac{<v_z>R^2}{4D_A}\left[\left(\frac{r}{R} \right)^2 \quad \frac{1}{2}\left(\frac{r}{R} \right)^4 - \frac{1}{3} \right] \tag{7.42}$$

so that, from Equation 7.38, we find

$$\tilde{c}_A = \frac{<v_z> R^2}{4D_A}\left[\left(\frac{r}{R} \right)^2 - \frac{1}{2}\left(\frac{r}{R} \right)^4 - \frac{1}{3} \right] \frac{\partial <c_A>}{\partial z} \tag{7.43}$$

We now can evaluate the dispersion term in Equation 7.19. Using Equations 7.31 and 7.43 we find, after manipulations,

$$-\frac{\partial <\tilde{v}_z \tilde{c}_A>}{\partial z} = \frac{<v_z>^2 \ R^2}{48 D_A} \frac{\partial^2 \ <c_A>}{\partial z^2} \quad (7.44)$$

Substitution into Equation 7.19 leads to

$$\frac{\partial <c_A>}{\partial t} + <v_z> \frac{\partial <c_A>}{\partial z} = \left(D_A + \frac{<v_z>^2 \ R^2}{48 D_A} \right) \frac{\partial^2 \ <c_A>}{\partial z^2} \quad (7.45)$$

This equation is usually expressed in terms of an effective diffusivity defined by

$$D_e = D_A + \frac{<v_z>^2 \ R^2}{48 D_A} \quad (7.46)$$

so that

$$\frac{\partial <c_A>}{\partial t} + <v_z> \frac{\partial <c_A>}{\partial z} = D_e \frac{\partial^2 \ <c_A>}{\partial z^2} \quad (7.47)$$

It is interesting to note that the governing equation for the average concentration looks like the convection–diffusion equation but replacing the diffusivity by the effective diffusivity.

7.2 GENERALIZATION OF THE DISPERSION MODEL

An interesting aspect of Equation 7.47 is that it is mathematically identical to the convection–diffusion equation, but it has two fundamental differences: first, the concentration represented by the equation is not a point value but rather an average; and second, the diffusivity has been replaced by an effective diffusivity that is the sum of the diffusivity plus a contribution from the effect of the local flow nonuniformities on the average concentration field. This contribution is sometimes called the **dispersion coefficient** or **dispersivity**,

$$D^* = \frac{<v_z>^2 \ R^2}{48 D_A} \quad (7.48)$$

Note that we can rewrite this equation as follows:

$$\frac{D^*}{D_A} = \frac{<v_z>^2 \ R^2}{48 D_A^2} = \frac{Pe^2}{48} \quad (7.49)$$

In most applications of interest, Pe is high enough that the dispersion coefficient becomes the dominant term in Equation 7.47.

The Taylor dispersion analysis can be generalized to other situations in which the dependent variable of interest is an average concentration. Applications include

1. Flow through porous media (see Chapter 8). In this case the concentration to be modeled represents a spatial average concentration defined at the pore scale where velocity fluctuations occur over distances shorter than the size of one pore. The averaged continuity equation is similar to the convection–diffusion equation,

$$\frac{\partial <c_A>}{\partial t} + <\mathbf{v}_z> \cdot\nabla <c_A> = \nabla \cdot (\mathbf{D}_e \cdot \nabla <c_A>) \tag{7.50}$$

   with the exception that the effective diffusivity is a tensor (since velocity nonuniformities might depend on direction).
2. Turbulent flows. The average concentration in this case is a time-averaged concentration ($\bar{c}_A$) and the continuity equation becomes

$$\frac{\partial \bar{c}_A}{\partial t} + \bar{\mathbf{v}}_z \cdot \nabla \bar{c}_A = \nabla \cdot (\mathbf{D}_e \cdot \nabla \bar{c}_A) \tag{7.51}$$

Here, the tensor $\mathbf{D}_e$ is the **turbulent dispersivity**.

The Taylor–Aris theory presented before to describe the transport of solute in a pipe under laminar flow conditions illustrates aspects of the general mechanism of transport called dispersion. In general, dispersion is characterized by

1. It is a pseudo-transport mechanism that governs transport when the flow is represented as a mean flow (in terms of an average velocity) and the concentration of transported species is represented by an average. In Taylor–Aris dispersion, the averaging process involves cross-sectional averages of the velocity and concentration.
2. The cause of dispersive transport is a fluctuation of the velocity field. In Taylor–Aris dispersion, the radial variation of point velocity causes the spread in the distribution of transported species that is mathematically represented by the dispersion term.
3. Dispersion is "Fickian." This means that dispersive transport can be interpreted as a flux that has the same mathematical form as Fick's law (more on this below). As seen before, the Taylor–Aris theory leads to the following average transport equation for laminar flow in a pipe

$$\frac{\partial \bar{c}}{\partial t} + \bar{v}_z \frac{\partial \bar{c}}{\partial z} = \bar{D}^* \frac{\partial^2 \bar{c}}{\partial z^2} \tag{7.52}$$

where the dispersivity (or dispersion coefficient) is given by

$$\bar{D}^* = D + \frac{R^2 \bar{v}_z^2}{48D} \tag{7.53}$$

The Fickian term containing the coefficient given by the second term on the right-hand side of this equation represents the contribution of dispersion to transport.

Any other system in which transport is quantified by average velocity and concentration, and in which the velocity exhibits fluctuations with respect to its average, can be represented by the dispersion model, replacing point variables with averaged variables, and diffusivity with dispersivity.

7.3 TURBULENT DISPERSION

Spatial fluctuations of velocity are not the only cause of hydrodynamic dispersion. Temporal fluctuations also give rise to a dispersive transport mechanism. It is known that turbulent flows, which occur widely in practice, represent an unstable type of flow whose structure is characterized by

1. The velocity vector is fully three-dimensional.
2. Components of the velocity fluctuate continuously with time, with fluctuation periods of the order of 1 ms.
3. The velocity fluctuations appear random.

For example, Figure 7.2 illustrates how the magnitude of the fluid velocity changes with time in a turbulent flow. The fluctuations are so rapid that sometimes even the measuring device cannot keep up with them and will record a time-smoothed value.

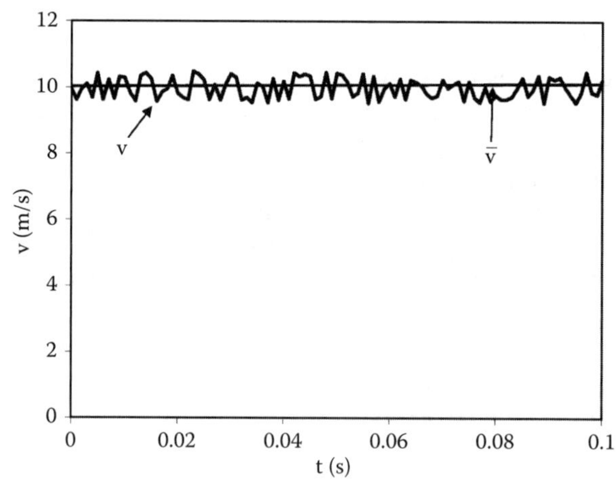

FIGURE 7.2 Typical velocity fluctuations in turbulent flow. In this example, even though the velocity changes with time, there is an average value of the velocity that remains constant.

The analysis of turbulent flows is done in terms of time-averaged variables. The time-averaged velocity vector is defined by

$$\overline{\mathbf{v}} = \frac{1}{2\Delta t}\int_{t-\Delta t}^{t+\Delta t} \mathbf{v}dt \tag{7.54}$$

where Δt is a time period such that $\Delta t \gg \Delta t_f$, where Δt_f is a characteristic time period for the fluctuations, $\Delta t_f = O(1 \text{ ms})$. The constraint specified ensures that the average velocity will be a smooth function of time. In the example in Figure 7.2, this function is the constant function.

Similarly, we define a time-averaged concentration by

$$\overline{c} = \frac{1}{2\Delta t}\int_{t-\Delta t}^{t+\Delta t} c dt \tag{7.55}$$

(in the rest of this analysis, the overbar will represent time averages).

On the basis of the previous definitions, we can define the fluctuation velocity and concentration by

$$\mathbf{v}' = \mathbf{v} - \overline{\mathbf{v}} \tag{7.56}$$

$$c' = c - \overline{c} \tag{7.57}$$

Note that v′ and c′ represent the fluctuations of velocity and concentration around their averages.

The analysis of turbulent flows starts with the point continuity equation which, for dilute systems with constant diffusivity, is, in the absence of chemical reactions,

$$\frac{\partial c}{\partial t} + \mathbf{v}\cdot\nabla c = D\nabla^2 c \tag{7.58}$$

This equation is time averaged:

$$\frac{1}{2\Delta t}\int_{t-\Delta t}^{t+\Delta t}\left(\frac{\partial c}{\partial t} + \mathbf{v}\cdot\nabla c\right)dt = \frac{1}{2\Delta t}\int_{t-\Delta t}^{t+\Delta t} D\nabla^2 c\, dt \tag{7.59}$$

After manipulations (see, e.g., Clark, 1996), this equation becomes

$$\frac{\partial \overline{c}}{\partial t} + \overline{\mathbf{v}}\cdot\nabla\overline{c} = D\nabla^2\overline{c} - \nabla\cdot\overline{(\mathbf{v}'c')} \tag{7.60}$$

Note that the only mathematical difference between this equation and the original continuity equation is the last term, which will represent dispersive transport.

The representation of the process in terms of averaged variables leads to a loss of information with regard to what happens over very short times and distances. For this reason, a representation for the dispersive term in Equation 7.60 could be sought empirically; that is, the representation for the dispersive transport term will be a sort of constitutive equation, which resembles the formulation of the diffusive flux in terms of concentration gradients using Fick's law. In fact, the quantity $\overline{\mathbf{v}'c'}$ is called "dispersive flux," and is usually quantified empirically. For the case of flow in a pipe, experimental evidence shows that the dispersive flux can be represented as follows:

$$\overline{\mathbf{v}'c'} = -D_t \nabla \bar{c} \tag{7.61}$$

where the coefficient D_t is called turbulent dispersivity or eddy dispersivity. It is not surprising that the dispersive flux has this form (which closely resembles Fick's law) since the time-averaging process and the transport due to local fluctuations is very similar to the random motion of molecules and the transport mechanism that this induces (diffusion). For developed pipe flow, the only nonzero component of the average velocity is the z component, and, considering that D_t is uniform, Equation 7.9 simplifies to

$$\frac{\partial \bar{c}}{\partial t} + \bar{v}_z \frac{\partial \bar{c}}{\partial z} = (D + D_t)\frac{\partial^2 \bar{c}}{\partial z^2} \tag{7.62}$$

where we have assumed that the average concentration does not change with radial position. An interesting aspect of this equation is that it is mathematically identical to the equation obtained in Taylor–Aris theory. In fact, with almost no exceptions, it will happen that $D_t \gg D$, and the diffusivity is usually omitted from Equation 7.62. The turbulent dispersivity is correlated empirically to experimental data. One such correlation for turbulent flow in a pipe is

$$\frac{D_t}{\bar{v}_z (2R)} = \frac{3 \times 10^7}{Re^{2.1}} + \frac{1.35}{Re^{1/8}}, \; Re > 2100 \tag{7.63}$$

where the Reynolds number is defined by

$$Re = \frac{\rho \bar{v}_z 2R}{\mu} \tag{7.64}$$

where ρ and μ are the density and viscosity of the fluid, respectively.

Taking into account that Equations 7.47 and 7.62 are mathematically similar, we can think of a continuous representation for the dispersivity from laminar to turbulent flow. To find this representation, we will use as independent variable the Peclet number, defined by

$$Pe = \frac{\bar{v}_z 2R}{D} \tag{7.65}$$

Rewriting Equation 7.53 in terms of the Peclet number yields

$$\frac{\bar{D}^*}{D} = 1 + \frac{Pe^2}{192} \tag{7.66}$$

The Peclet number also can be represented as

$$Pe = ReSc \tag{7.67}$$

where the Schmidt number is defined by

$$Sc = \frac{\mu}{\rho D} \tag{7.68}$$

Equation 7.63 now can be rewritten as follows:

$$\frac{D_t}{D} = \frac{3 \times 10^7 Sc}{Re^{1.1}} + 1.35 Re^{7/8} Sc \tag{7.69}$$

Figure 7.3 shows a comparison on dimensionless dispersivities between laminar and turbulent flows. The drop in the value of the dispersivity as the flow regime changes is due to the relatively flat velocity profile obtained in turbulent flow, which yields a somewhat lower degree of dispersion from the laminar flow case at the same Reynolds number.

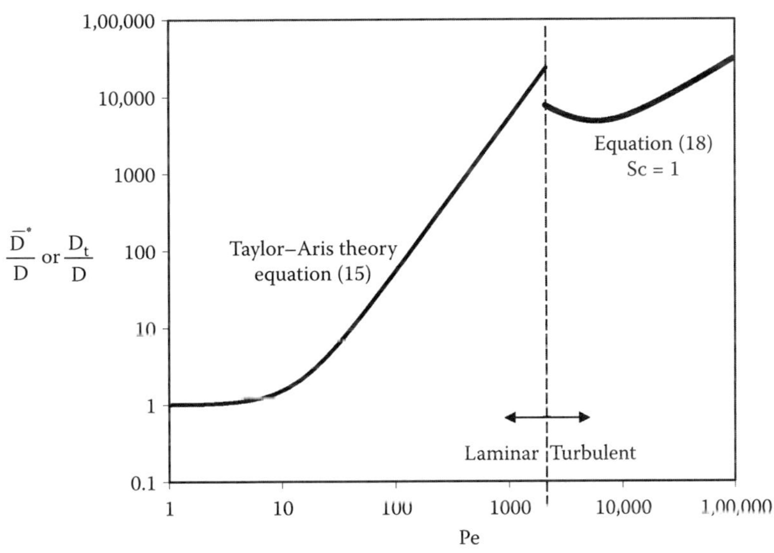

FIGURE 7.3 Dispersion coefficients for flow in a pipe.

In a general, three-dimensional flow, the dispersive flux requires a representation that is more general than Equation 7.61:

$$\overline{\mathbf{v}'c'} = -\mathbf{D}\cdot\nabla\bar{c} \tag{7.70}$$

where **D** is the **dispersion tensor**, which in Cartesian coordinates can be represented as follows:

$$\mathbf{D} = \begin{bmatrix} D_{xx} & 0 & 0 \\ 0 & D_{yy} & 0 \\ 0 & 0 & D_{zz} \end{bmatrix} \tag{7.71}$$

The diagonal form of the tensor is valid when the coordinate system is aligned with the principal axes of dispersion. Substitution of Equation 7.70 into Equation 7.60 leads to

$$\frac{\partial c}{\partial t} + \mathbf{v}\cdot\nabla c = D_{xx}\frac{\partial^2 c}{\partial x^2} + D_{yy}\frac{\partial^2 c}{\partial y^2} + D_{zz}\frac{\partial^2 c}{\partial z^2} \tag{7.72}$$

where we have considered that the dispersivities are constant, and we have dropped the overbars to simplify notation. Some analytical solutions for this equation are presented below.

7.4 SOLUTIONS OF THE CONVECTIVE–DISPERSION EQUATION

Motion in the x direction only with velocity u, constant dispersivities:

$$\frac{\partial c}{\partial t} + u\frac{\partial c}{\partial x} = D_{xx}\frac{\partial^2 c}{\partial x^2} + D_{yy}\frac{\partial^2 c}{\partial y^2} + D_{zz}\frac{\partial^2 c}{\partial z^2}$$

1. Instantaneous point source (N moles) at $x = y = z = 0$ and $t = 0$, unbounded medium

$$c = \frac{N}{(2\pi)^{3/2}(P_{xx}P_{yy}P_{zz})^{1/2}}\exp\left[-\frac{(x-ut)^2}{2P_{xx}} - \frac{y^2}{2P_{yy}} - \frac{z^2}{2P_{zz}}\right]$$

where

$$P_{xx} = 2D_{xx}t,\ P_{yy} = 2D_{yy}t,\ P_{zz} = 2D_{zz}t$$

2. Continuous point source ($\dot{N}$ moles per unit time) at x = y = z = 0, unbounded medium (steady-state solution)

$$c = \frac{\dot{N}}{4\pi(D_{yy}D_{zz}x^2 + D_{xx}D_{zz}y^2 + D_{yy}D_{xx}z^2)^{1/2}}$$
$$\exp\left\{-\frac{u}{2D_{xx}}\left[D_{xx}^{1/2}\left(\frac{x^2}{D_{xx}}+\frac{y^2}{D_{yy}}+\frac{z^2}{D_{zz}}\right)^{1/2}-x\right]\right\}$$

3. Continuous line source ($\dot{n}$ moles per unit time and length) on the y axis, unbounded medium (steady-state solution)

$$c = \frac{\dot{n}}{2\pi(D_{xx}D_{zz})^{1/2}}\exp\left(\frac{ux}{2D_{xx}}\right)K_0\left[\frac{u}{2}\left(\frac{x^2}{D_{xx}^2}+\frac{z^2}{D_{xx}D_{zz}}\right)^{1/2}\right]$$

where K_0 is the modified Bessel function of the second kind and zeroth order.

For the case ux » D_{xx} and x » z, this solution simplifies to

$$c = \frac{\dot{n}}{2(\pi D_{zz}ux)^{1/2}}\exp\left(-\frac{uz^2}{4D_{zz}}x\right)$$

4. Continuous point source ($\dot{N}$ moles per unit time) at x = 0, y = 0, z = h, with no-flux condition on the plane z = 0, and D_{xx} = 0 (steady-state solution)

$$c = \frac{\dot{N}}{4\pi x(D_{yy}D_{zz})^{1/2}}\exp\left(-\frac{uy^2}{4xD_{yy}}\right)\left\{\exp\left[-\frac{u(z-h)^2}{4xD_{zz}}\right]+\exp\left[-\frac{u(z+h)^2}{4xD_{zz}}\right]\right\}$$

5. Continuous line source ($\dot{n}$ moles per unit time and length) at x = 0, z = h, with no-flux condition on the plane z = 0, and D_{xx} = 0 (steady-state solution)

$$c = \frac{\dot{n}}{2(\pi uxD_{zz})^{1/2}}\left\{\exp\left[-\frac{u(z-h)^2}{4xD_{zz}}\right]+\exp\left[-\frac{u(z+h)^2}{4xD_{zz}}\right]\right\}$$

As an example of application of the equations above, consider a continuous point source in an unbounded medium. The following values are used in the calculation: N = 1, u = 1, $D_{xx} = D_{yy} = D_{zz} = 1$. The concentration profile was generated by the appropriate equation (item 2 in the list above). Figures 7.4 and 7.5 show plots of the concentration profiles around the source. The coordinate system in these figures has been translated so that the point source is located at the center of the x–y plane represented in the figures.

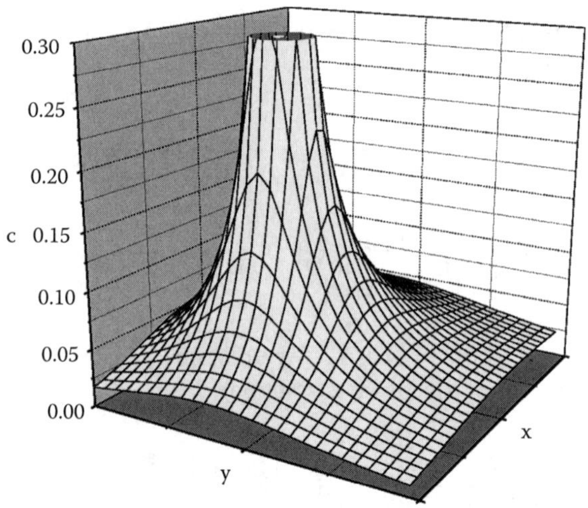

FIGURE 7.4 Steady-state concentration distribution around a continuous point source.

Even though this example is an abstract solution for specific numerical values of the parameters used only to show the trends represented by the equations, we can make an analogy with the discharge of a contaminant in the atmosphere from a smokestack. The scale of observation in Figures 7.4 and 7.5 is then large enough that the smokestack can be represented as a point, and the plots would correspond to the contaminant distribution over the plane on which the discharge occurs (i.e., at the elevation of the smokestack).

The results clearly show the effects of hydrodynamic dispersion: if there were no dispersion and only convection in the x direction, the contaminant would simply follow a straight line in the x direction and any lateral spread would be due to diffusion

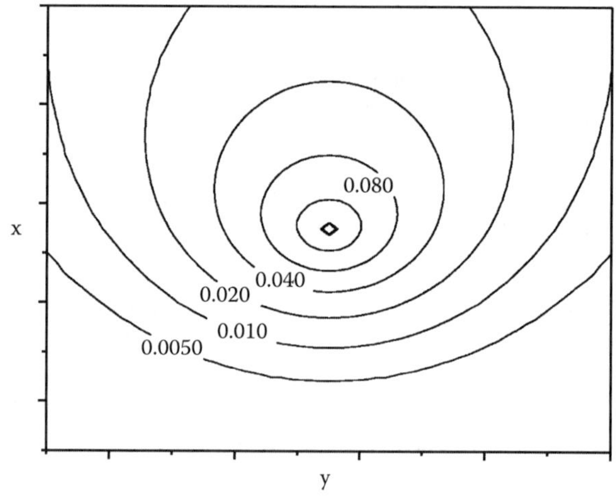

FIGURE 7.5 Concentration contours around a continuous point source.

(which would be many orders of magnitude slower than dispersion). Figures 7.4 and 7.5 show that, even though the contaminant is preferentially transported along the x direction (as evidenced by the higher concentrations directly downstream from the source), there is appreciable spread around the source, including upstream of the source.

PROBLEMS

7.1 We would like to assess the importance of dispersion in laminar pipe flow. Consider the steady-state dispersion of a solute (A) that decomposes according to a first-order chemical reaction. The convective–dispersion equation is

$$\langle v_z \rangle \frac{d \langle c_A \rangle}{dz} = D_e \frac{d^2 \langle c_A \rangle}{dz^2} - k \langle c_A \rangle$$

Appropriate boundary conditions are

$$\langle c_A \rangle = c_{A0}, z = 0$$

$$\langle c_A \rangle = 0, z \rightarrow \infty$$

with the effective diffusivity, D_e, given by Equation 7.46.

a. Write the problem in dimensionless form using the following definitions:

$$u = \frac{\langle c_A \rangle}{c_{A0}}$$

$$\xi = \frac{z}{R}$$

$$Pe = \frac{\langle v_z \rangle R}{D_A}$$

$$\phi = \frac{kR^2}{D_A}$$

Find $u = u(\zeta)$.

b. Plot u vs. ξ for the case $\phi = 100$, $Pe = 500$. In the same graph, plot u vs. ξ for the same conditions, but assuming that $D_e = D_A$ (this would be the hypothetical case in which there is no dispersion).

7.2 The effluent from a wastewater treatment plant is discharged to a river. The plant fails to remove a trace organic contaminant of concern (A) that is originally absent from the river, and the discharge rate is such that, at the point of discharge ($z = 0$, where z is longitudinal distance along the river in the direction of motion), the cross-sectional average concentration of contaminant is $<c_A>_0$. The contaminant sorbs to the sediments at the bottom of the river via a first-order process and is subsequently biodegraded. The convection–dispersion equation is

$$\frac{\partial <c_A>}{\partial t} + <v_z> \frac{\partial <c_A>}{\partial z} = D_e \frac{\partial^2 <c_A>}{\partial z^2} - k_e <c_A>$$

where the effective diffusivity, D_e, the effective first-order rate constant, k_e, and the water average velocity, $<v_z>$, are known.

a. Find $<c_A>(z)$ at **steady state** both upstream ($z < 0$) and downstream ($z > 0$) of the discharge point.
b. Levels of contaminant higher than $0.001 <c_A>_0$ are not allowed in drinking water. Find an expression to calculate the distances downstream (L_d) and upstream (L_u) from the discharge point beyond which it is safe to use the river as a source of drinking water.

8 Transport in Multiphase Systems

8.1 MULTIPHASE SYSTEMS AND POROUS MEDIA

Multiphase systems and porous media are ubiquitous in environmental engineering applications. Here, we will use the term **porous media** to describe multiphase systems consisting of a stationary solid phase and one or more, stationary or moving, fluid phases that occupy the interstitial space left void by the solid phase. Examples of relevant porous media are groundwater aquifers, soils in general, porous membranes used in separations, and all types of beds packed with particles, such as those used in adsorption and ion exchange, catalytic reactors, trickling filters, and chromatography.

One of the most important aspects of porous media, which has a bearing on the description of transport processes, is the description of their geometric structure. In general, we are not interested in producing a mapping of the distribution of the phases point by point, but rather we seek a description based on properties defined over a scale bigger than the typical pore size (more on this below).

An important structural property of porous media is the porosity. To define it, consider a typical cross section of a porous medium (Figure 8.1). We will define control volumes that will be spheres of volume V whose center is a specific point in the medium such that that point falls in the interstitial (fluid) space. For each volume, we will calculate the ratio of fluid volume (V_f) to total volume:

$$\xi = \frac{V_f}{V} \tag{8.1}$$

For small values of V, the ratio will be 1 since the volume inside the sphere is only occupied by fluid. As the increasing volume touches the first solid particle, then $V = V_f + V_s$, where V_s is the volume occupied by the solid phase, and the ratio will become lower than 1. As the scale of V is increased, oscillations in the value of ξ are measured, but, at large enough scales, these oscillations dampen until the measured value of ξ becomes independent of V. This constant value is the porosity of the medium (π).

When V is large enough that the volume ratio becomes uniform, we operate on a scale that is large compared with the size of the solid particles (or, in general, the size of the pores), and the porous medium can be taken to be a continuum. We will assume that this scale is still much smaller than the scale over which we would like to describe

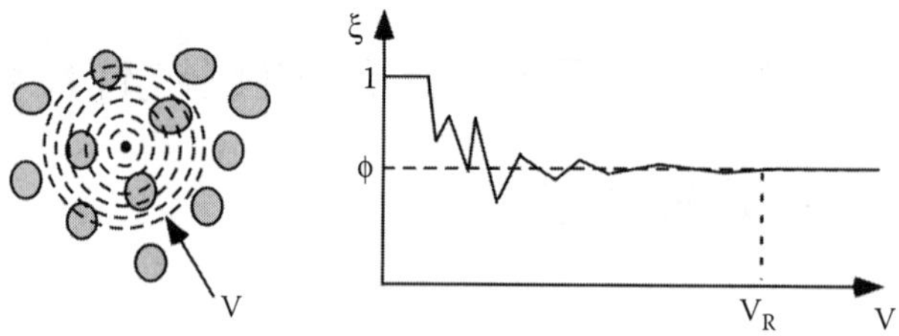

FIGURE 8.1 Ratio of fluid (void) volume in a porous medium to total volume, as the scale of observation is enlarged from a point.

transport processes, so that a "point" in our description scale can be thought of as including a relatively large number of particles (or pores). The volume for which the porous medium becomes a continuum (V_R) is called "representative volume" and in packed beds it usually contains 10–50 particles. A proper definition of porosity will be

$$\pi = \xi, \text{ for } V > V_R \tag{8.2}$$

We will represent transport at scales $V > V_R$, using variables that represent averages over the representative volume of porous medium. Therefore, our description will not distinguish between fluid and solid phases from a structural point of view, but rather it will consider the porous medium as a continuum. The definition of average quantities will be based on spatial averages. For a property ψ defined at each point in the porous medium, we define the spatial average by

$$<\psi> = \frac{1}{V_R}\int_{V_R} \psi dV \tag{8.3}$$

for example, consider the function that is unity in the fluid phase and zero in the solid phase:

$$\psi = \begin{cases} 1 \text{ in } V_f \\ 0 \text{ in } V_s \end{cases} \tag{8.4}$$

Then,

$$<\psi> = \frac{1}{V_R}\int_{V_f} dV = \frac{V_f}{V_R} = \phi \tag{8.5}$$

that is, the average of the function is the porosity.

We can use Equation 8.3 to define an average velocity:

$$\langle \mathbf{v} \rangle = \frac{1}{V_R} \int_{V_R} \mathbf{v}\, dV \tag{8.6}$$

Since the solid is stationary, $\mathbf{v} = 0$ in V_s and the average velocity contains an integral of the fluid velocity only,

$$\langle \mathbf{v} \rangle = \frac{1}{V_R} \int_{V_f} \mathbf{v}\, dV \tag{8.7}$$

This average velocity is called **superficial velocity**. For one-dimensional flows, the superficial velocity represents the volumetric flow rate of fluid per unit cross-sectional area of porous medium: $Q = \langle v \rangle A$, where A is the cross-sectional area of medium normal to the direction of motion.

A characteristic of the superficial velocity is that it does not represent an average of the fluid velocity in the pore space. A proper definition for the average fluid velocity would be

$$\langle \mathbf{u} \rangle = \frac{1}{V_f} \int_{V_f} \mathbf{v}\, dV \tag{8.8}$$

This is called the **interstitial velocity** or intrinsic-average velocity. Comparing Equations 8.7 and 8.8 we can see that

$$\langle \mathbf{u} \rangle = \frac{\langle \mathbf{v} \rangle}{\phi} \tag{8.9}$$

Both superficial and interstitial velocities are used to represent flow through porous media.

Other physically relevant variables also are represented in terms of averages. For example, in the study of transport of a chemical species in the fluid phase of a porous medium when the species is insoluble in the solid phase, an interstitial average of the concentration is preferred,

$$\langle c \rangle = \frac{1}{V_f} \int_{V_f} c\, dV \tag{8.10}$$

The interstitial average has the property that if c is uniform in the fluid phase, then $\langle c \rangle = c$.

In what follows, to avoid the use of complex notation, we will drop the angular brackets to represent averages and use the following equivalences (see Figures 8.2 and 8.3 and Table 8.1):

<v> ⇒ **v** (superficial velocity)
<u> ⇒ **u** (interstitial velocity)
<c> = c (interstitial average concentration)

(a)

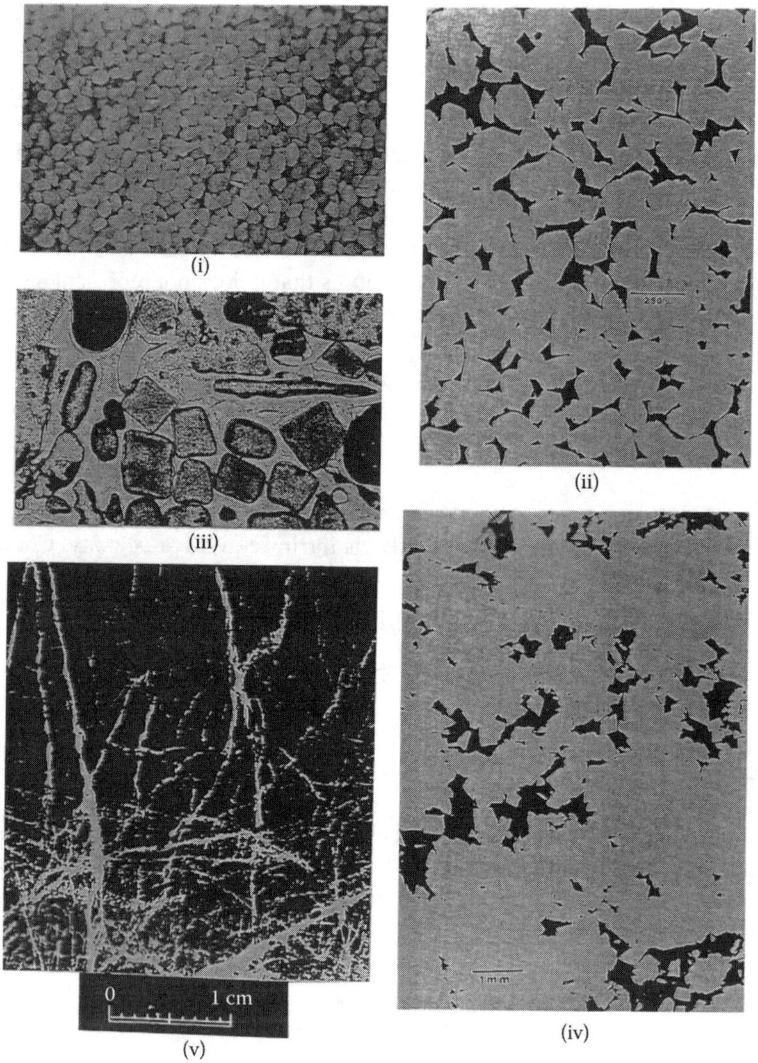

FIGURE 8.2 Examples of porous media microstructure: (i) sand, (ii) sandstone, (iii) limestone, (iv) dolomitized limestone, and (v) chalk. (From Adler PM, *Porous Media. Geometry and Transports*, Butterworth-Heinemann, Boston, 1992.)

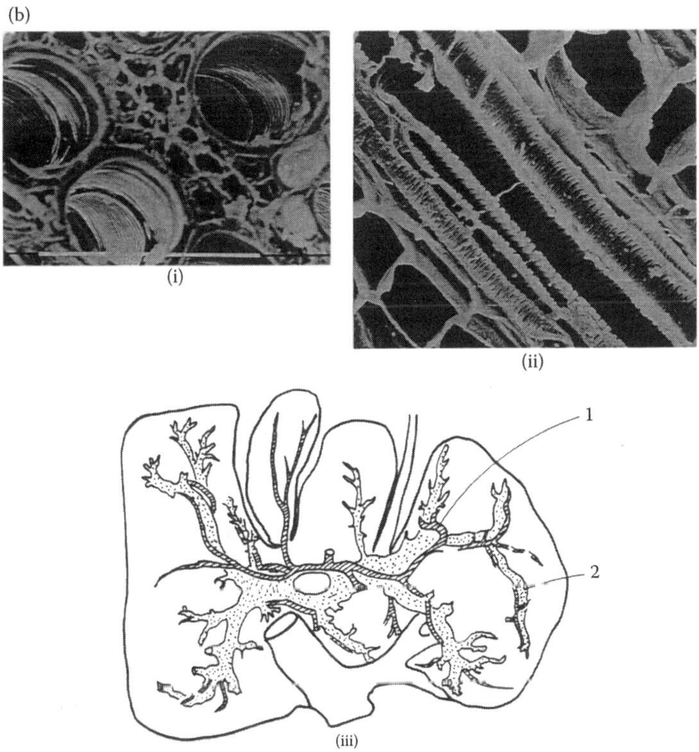

FIGURE 8.2 (continued) Examples of porous media microstructure: (i) transverse, (ii) longitudinal cross sections of capillaries in the petiole of a sunflower, and (iii) arteries (1) and veins (2) in the human liver. (From Adler PM, *Porous Media. Geometry and Transports*, Butterworth-Heinemann, Boston, 1992.)

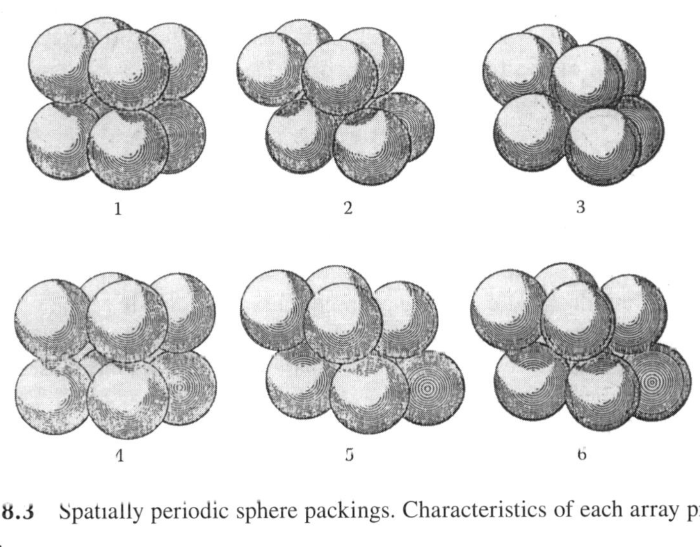

FIGURE 8.3 Spatially periodic sphere packings. Characteristics of each array provided in Table 8.1.

TABLE 8.1
Characteristic of Regular Sphere Packings

Case	Packing Name	Interlayer Spacing[a]	Coordination Number[b]	π
1	Cubic	2R	6	0.476
2	Orthorhombic	$\sqrt{3}$R	8	0.395
3	Hexagonal	$\sqrt{2}$R	12	0.259
4	Orthorhombic	2R	8	0.395
5	Tetragonal	$\sqrt{3}$R	10	0.302
6	Hexagonal	$\sqrt{8/3}$R	12	0.259

[a] Vertical distance between sphere centers.
[b] Number of spheres that touch any given sphere.

8.2 SINGLE-PHASE FLOW IN POROUS MEDIA: DARCY'S LAW

The motion of Newtonian fluids in incompressible flow through porous media has been extensively investigated in the literature both from theoretical and experimental points of view. The mathematical representations of the principles of mass and momentum conservation in terms of spatially averaged variables are well known for the case of relatively slow flow in the pores. A criterion for the application of the equations presented below is based on a characteristic Reynolds number at the pore scale, defined by

$$\mathrm{Re} = \frac{\rho v_p d_p}{\mu} \tag{8.11}$$

where v_p and d_p are a characteristic pore velocity and a characteristic pore length, respectively. For Re < 1, which the most typical case encountered in practical applications, the motion of the fluid is described by Darcy's law. For one-dimensional flow in the z direction, Darcy's law is

$$v_z = -\frac{K}{\mu}\left(\frac{dP}{dz} - \rho g_z\right) \tag{8.12}$$

In this equation, P is the intrinsic average of the pressure in the pore space and dP/dz is, therefore, the gradient of the average pressure in the direction of flow, g_z is the component of the acceleration of gravity vector in the direction of flow and K is the **permeability** of the porous medium, a parameter that is an exclusive function of the microstructure of the medium.

Sometimes, Equation 8.12 is written as follows:

$$v_z = -\alpha\frac{dh}{dz} \tag{8.13}$$

where h is called hydraulic head, and is defined by

$$h = \frac{P}{\rho g} - \frac{g_z}{g} z \tag{8.14}$$

and α is the hydraulic conductivity,

$$\alpha = \frac{K\rho g}{\mu} \tag{8.15}$$

Darcy's law indicates that driving forces for motion are pressure drop (–dP/dz) and gravitational forces. The most important characteristic of Equation 8.12 is the fact that all effects on motion of the pore-scale structure are contained in just one coefficient, the permeability. Experimental and theoretical studies have determined that the permeability is affected by two structural characteristics: the porosity and a measure of the characteristic pore size. For example, for porous media made up of disordered packings of monodisperse spherical particles, the relevant pore-scale characteristic length is the particle diameter. However, instead of using the particle diameter, a hydraulic diameter is used instead. The hydraulic diameter is defined by

$$d_h \sim \frac{V_f}{A_s} \tag{8.16}$$

where V_f is the fluid volume in a representative volume of porous medium, and A_s is the wetted area of solid phase in the representative volume. Equation 8.16 can be written as

$$d_h \sim \frac{V_s}{A_s}\frac{V_f}{V_s} = \frac{V_s}{A_s}\frac{V_f}{V_R - V_f} \tag{8.17}$$

For monodisperse spheres (radius R, diameter d = 2R), we have

$$\frac{V_s}{A_s} = \frac{\frac{4}{3}\pi R^3}{4\pi R^2} = \frac{R}{3} = \frac{d}{6} \tag{8.18}$$

Also,

$$\frac{V_f}{V_R - V_f} = \frac{V_f/V_R}{1 - V_f/V_R} = \frac{\phi}{1-\phi} \tag{8.19}$$

so that

$$d_h \sim \frac{d}{6}\frac{\phi}{1-\phi} \tag{8.20}$$

Dropping the coefficient, we can define the hydraulic diameter as

$$d_h = \frac{\phi d}{1-\phi} \tag{8.21}$$

To determine a correlation for the permeability in sphere packings, Equation 8.12 is written in terms of the interstitial velocity,

$$u_z = -\frac{1}{\mu}\frac{K}{\pi}\left(\frac{dP}{dz} - \rho g_z\right) \tag{8.22}$$

It is postulated that the structural parameter K/ϕ only depends on d_h. In fact, experimental data show that K/ϕ is directly proportional to d_h^2. Therefore, an empirical equation for the permeability of sphere packings is

$$K = \frac{d^2\phi^3}{A(1-\phi)^2} \tag{8.23}$$

where A is a dimensionless constant whose value ranges from 130 to 200, with A = 180 being the most representative value. Equation 8.23 is called the Carman–Kozeny equation.

For three-dimensional flows, the general expression of Darcy's law is related to the concept of isotropy with respect to the flow process. Isotropy is a concept related to the dependence of a specific phenomenon or parameter on the direction in which it is occurring. A porous medium is said to be isotropic with respect to the flow process if the imposition of a pressure (or hydraulic head) gradient in different directions leads to the same superficial velocity. Under these circumstances, the three-dimensional version of Darcy's law is

$$\mathbf{v} = -\frac{K}{\mu}(\nabla P - \rho\mathbf{g}), \text{ isotropic media} \tag{8.24}$$

Sphere packings are examples of isotropic porous media. In some cases, the medium structure has features that imply an orientation of the local geometry. This type of medium is said to be anisotropic with respect to the flow process, and Darcy's law is

$$\mathbf{v} = -\frac{1}{\mu}\mathbf{K}\cdot(\nabla P - \rho\mathbf{g}) \tag{8.25}$$

where the permeability, **K**, is now a second-order tensor. If the coordinate system is aligned with the directions of orientation of the microstructure, **K** is a diagonal tensor and the expanded form of Equation 8.26 becomes, in Cartesian coordinates,

$$v_x = -\frac{1}{\mu} K_{xx} \left(\frac{\partial P}{\partial x} - \rho g_x \right) \tag{8.26}$$

$$v_y = -\frac{1}{\mu} K_{yy} \left(\frac{\partial P}{\partial y} - \rho g_y \right) \tag{8.27}$$

$$v_z = -\frac{1}{\mu} K_{zz} \left(\frac{\partial P}{\partial z} - \rho g_z \right) \tag{8.28}$$

Examples of porous media that are anisotropic with respect to the flow process are: wood (because of fiber alignment), packings of elongated particles in which particles are all aligned in the same direction, and consolidated media in which the pores run in a particular direction.

8.3 DIFFUSION IN POROUS MEDIA AND MULTIPHASE SYSTEMS

We will consider the most common case of diffusion in porous media, namely, that a chemical species present at low concentration diffuses through the fluid phase while being insoluble in the solid. Under these conditions, the presence of solid phase will be merely an obstacle to diffusion pathways. The equation that governs the process at a point in the fluid is the diffusion equation. A macroscopic equation can be obtained by averaging the point equation over the interstitial space. If the porous medium is isotropic with respect to the diffusion process, and its microstructure is uniform (i.e., the porosity does not change with position in the medium) the average equation is

$$\frac{\partial c}{\partial t} = D_{eff} \nabla^2 c \tag{8.29}$$

where D_{eff} is the **effective diffusivity**. This equation is mathematically identical to the point diffusion equation, with the differences that c now represents intrinsic average concentration and the diffusion coefficient is replaced by the effective diffusivity.

The effective diffusivity is lower than the molecular diffusivity ($D_{eff} < D$), mainly because of the tortuosity of the diffusion paths that molecules must follow (Figure 8.4), for which the solid phase acts as a physical obstacle.

Effective diffusivities must be measured for a particular porous medium as a function of microstructure. It is known that the main microstructural parameter that affects the effective diffusivity is the medium porosity. For example, Figure 8.5 shows experimental data and some theoretical prediction of effective diffusivities in isotropic particle packings.

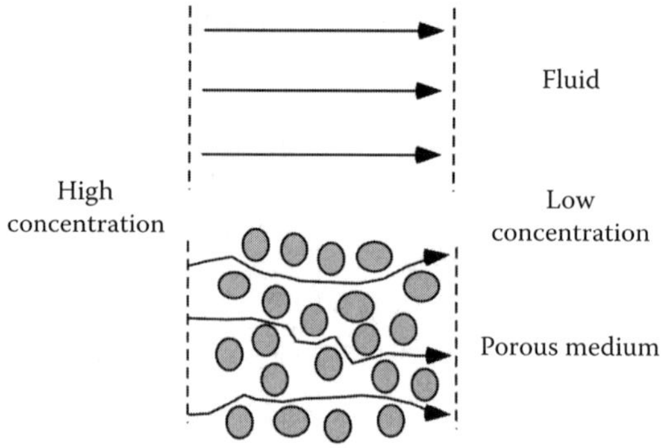

FIGURE 8.4 Diffusion paths in a homogeneous fluid and a porous medium.

For the case of media that are anisotropic with respect to the diffusion process (media with geometrical anisotropy) such as those that lead to the anisotropic version of Darcy's law, the generalized diffusion equation is

$$\frac{\partial c}{\partial t} = \nabla \cdot (\mathbf{D}_{\text{eff}} \cdot \nabla c) \tag{8.30}$$

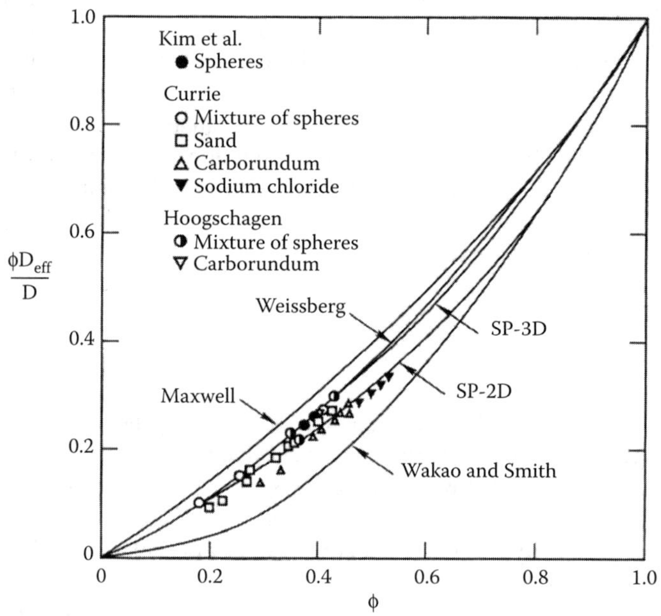

FIGURE 8.5 Effective diffusivities for packed beds. Symbols represent experimental data; Maxwell and Weissberg represent theoretical upper bounds; SP-2D and SP-3D represent theoretical results for cubic arrays of cylinders and spheres, respectively. (From Whitaker S, *The Method of Volume Averaging*, Kluwer Academic Publishers, Dordrecht, the Netherlands, 1999, and references therein.)

where $\mathbf{D}_{eff}$ is the **effective diffusivity tensor**. If, again, the coordinate axes are oriented with the main directions of anisotropy, the effective diffusivity tensor becomes a diagonal tensor and Equation 8.30 can be written as (also considering uniform effective diffusivities)

$$\frac{\partial c}{\partial t} = D_{xx}\frac{\partial^2 c}{\partial x^2} + D_{yy}\frac{\partial^2 c}{\partial y^2} + D_{zz}\frac{\partial^2 c}{\partial z^2} \tag{8.31}$$

8.4 DISPERSION IN POROUS MEDIA

The analysis of convective mass transport in porous media is based on the application of the convective–diffusion equation in the fluid phase at the pore level. In the absence of chemical reactions and considering that the solid phase is inert (i.e., the species analyzed does not adsorb on or dissolve in the solid particles), averaging of this equation over a representative volume of porous medium leads to the following macroscopic equation:

$$\frac{\partial c}{\partial t} + \mathbf{u}\cdot\nabla c = \nabla\cdot(\mathbf{D}^*\cdot\nabla c) \tag{8.32}$$

This equation is known as the convective–dispersion (or advective–dispersion) equation. The transport process considered is exclusively governed by diffusion and convection, and is called **passive dispersion**. The dispersivity tensor, $\mathbf{D}^*$, reflects the local velocity variations that occur at the pore level, in the same way that the Taylor–Aris dispersivity reflected the variations of axial velocity with radial position in a pipe. In this case, however, the mechanism is more complicated since the velocity vector at the pore level is three-dimensional. The main consequence of this is the fact that the dispersivity is a tensor, regardless of the local structure of the porous medium. For example, let us consider the case of average flow in the x direction: $\mathbf{u} = u\mathbf{e}_x$. Equation 8.32 can be written as

$$\frac{\partial c}{\partial t} + u\frac{\partial c}{\partial x} = \nabla\cdot(\mathbf{D}^*\cdot\nabla c) \tag{8.33}$$

or, in expanded form,

$$\frac{\partial c}{\partial t} + u\frac{\partial c}{\partial x} = \frac{\partial}{\partial x}\left(D^*_{xx}\frac{\partial c}{\partial x}\right) + \frac{\partial}{\partial y}\left(D^*_{yy}\frac{\partial c}{\partial y}\right) + \frac{\partial}{\partial z}\left(D^*_{zz}\frac{\partial c}{\partial z}\right) \tag{8.34}$$

Dispersive transport in the x direction (longitudinal dispersion) in this case is mainly due to the velocity profiles at the pore level: the velocity must drop from a certain maximum value to zero on the surface of the solid. This is a mechanism similar to Taylor–Aris dispersion. On the other hand, transport in directions perpendicular to the flow, y and z (lateral dispersion), is mainly a consequence of the three-dimensional nature of the local velocity due to the fact that the fluid must move laterally to flow around the solid particles. This induces mixing in the directions perpendicular to the flow. For an isotropic porous medium, one would expect the dispersivities in the lateral directions to be the same due to symmetry and also to have a lower value than the dispersivity in the direction of flow; that is,

$$D^*_{yy} = D^*_{zz} = D^*_T \tag{8.35}$$

$$D^*_{xx} = D^*_L \geq D^*_T \tag{8.36}$$

where D^*_L is the longitudinal dispersivity and D^*_T is the lateral dispersivity. For uniform, one-dimensional flow, Equation 8.34 is expressed as follows:

$$\frac{\partial c}{\partial t} + u\frac{\partial c}{\partial x} = D^*_L \frac{\partial^2 c}{\partial x^2} + D^*_T \frac{\partial^2 c}{\partial y^2} + D^*_T \frac{\partial^2 c}{\partial z^2} \tag{8.37}$$

A dimensional analysis of the point equations shows that, for an isotropic porous medium with uniform structure, the dispersivities depend on

1. The molecular diffusivity
2. Local structure of the porous medium (mainly on porosity)
3. A local Peclet number, defined in terms of the hydraulic diameter

$$Pe_p = \frac{ud_h}{D} \tag{8.38}$$

For monodisperse sphere packings, using Equation 8.20 leads to

$$Pe_p = \frac{ud\phi}{D(1-\phi)} \tag{8.39}$$

For packed beds, correlations have been developed in the form (see below)

$$\frac{D^*_L}{D} = \frac{D^*_L}{D}(Pe_p) \tag{8.40}$$

$$\frac{D^*_T}{D} = \frac{D^*_T}{D}(Pe_p) \tag{8.41}$$

In what follows we analyze experimental methods used to determine the longitudinal and lateral dispersivities as examples of passive dispersion processes.

8.4.1 Experimental Determination of the Longitudinal Dispersivity in 1-D Flows

In this experiment, a packed bed is filled with a liquid that initially contains a zero concentration of the component that will be used in the study of passive dispersion. At $t = 0$, the entering flow of liquid will have a constant concentration c_0. Concentrations of the dispersing species will be measured as a function of time at various axial positions along the bed (Figure 8.6). For example, the dispersing species could be an electrolyte (e.g., sodium chloride), whose concentration is directly proportional to the electrical conductivity of the liquid. In this case, the measuring probes are electrical conductivity probes. Alternatively, the sampling probes could be tubes through which liquid samples are extracted for analysis. If the volumetric flow rate of fluid entering the bed is Q and A is the cross-sectional area, the interstitial velocity is

$$u = \frac{v}{\phi} = \frac{Q}{A\phi} \tag{8.42}$$

In this experiment, there is nothing that would induce a lateral concentration gradient, so that $c = c(t,x)$. Equation 8.37 simplifies to

$$\frac{\partial c}{\partial t} + u\frac{\partial c}{\partial x} = D_L^* \frac{\partial^2 c}{\partial x^2} \tag{8.43}$$

Initial and boundary conditions are

$$c = 0, \ t = 0 \tag{8.44}$$

$$c = c_0, \ x = 0 \tag{8.45}$$

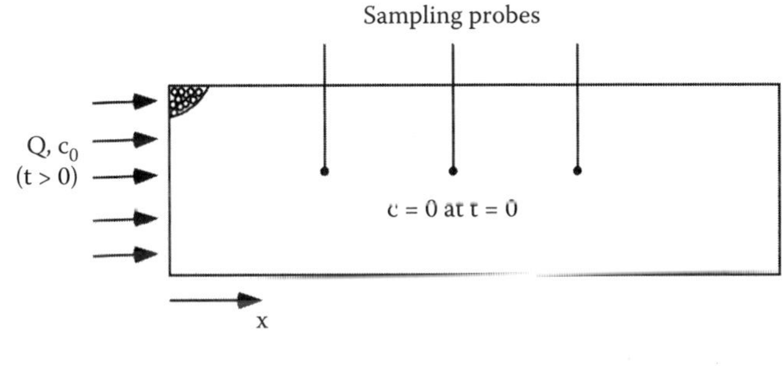

FIGURE 8.6 Experiment to determine the longitudinal dispersivity in a packed bed: a volumetric flow rate Q of liquid with a concentration c_0 flows continuously into the bed for $t > 0$. The concentration of dispersing species is measured as a function of x and t at sampling points.

$$c = 0, x \to \infty \tag{8.46}$$

It has been assumed that the bed is long enough downstream of the sampling probes that condition 8.47 applies.

The solution of the problem posed by Equations 8.43 through 8.46 is (the problem can be solved analytically using Laplace transform)

$$\frac{c}{c_0} = \frac{1}{2}\text{erfc}\left(\frac{x - ut}{2\sqrt{D_L^* t}}\right) + \frac{1}{2}\exp\left(\frac{ux}{D_L^*}\right)\text{erfc}\left(\frac{x + ut}{2\sqrt{D_L^* t}}\right) \tag{8.47}$$

This equation can be fitted to experimental data on c(t,x), using the longitudinal dispersivity as the only adjustable parameter.

Equation 8.47 can be expressed in dimensionless form as follows:

$$\frac{c}{c_0} = \frac{1}{2}\text{erfc}\left(\frac{X - T}{2\sqrt{T}}\right) + \frac{1}{2}\exp(X)\text{erfc}\left(\frac{X + T}{2\sqrt{T}}\right) \tag{8.48}$$

where

$$X = \frac{ux}{D_L^*} \tag{8.49}$$

and

$$T = \frac{u^2 t}{D_L^*} \tag{8.50}$$

A plot of Equation 8.48 is shown in Figure 8.7.

Figure 8.8 shows experimentally determined longitudinal dispersivities for beds of spheres. The experimental results are compared with Taylor–Aris theory, and theoretical results for arrays of cylinders. At low Pe_p, the longitudinal, dispersivity approaches the effective diffusivity. For $Pe_p > 1$, the trend is approximately linear in the log–log scale and follows the proportionality

$$\frac{D_L^*}{D} \sim Pe_p^{1.2} \tag{8.51}$$

8.4.2 Experimental Determination of the Lateral Dispersivity in 1-D Flows

For this experiment, a lateral concentration profile is created in the packed bed by dividing the flow at the entrance in two sections: one with a zero concentration,

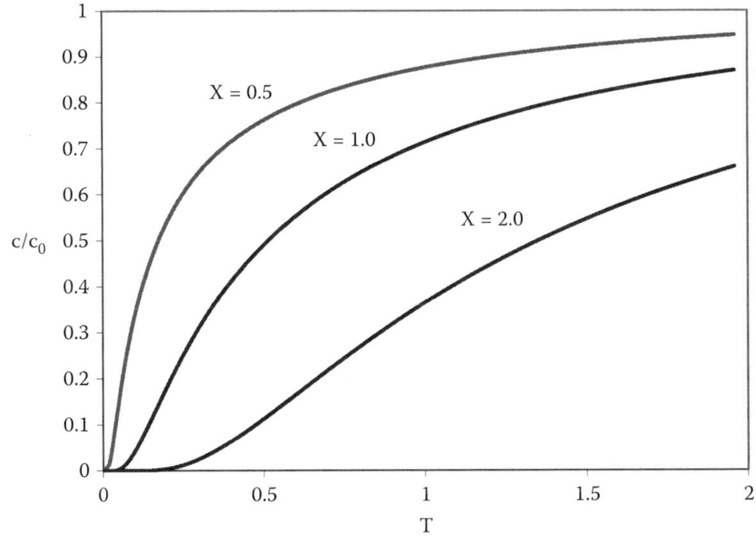

FIGURE 8.7 Dimensionless breakthrough curves: dimensionless concentration vs. dimensionless times at various dimensionless axial positions.

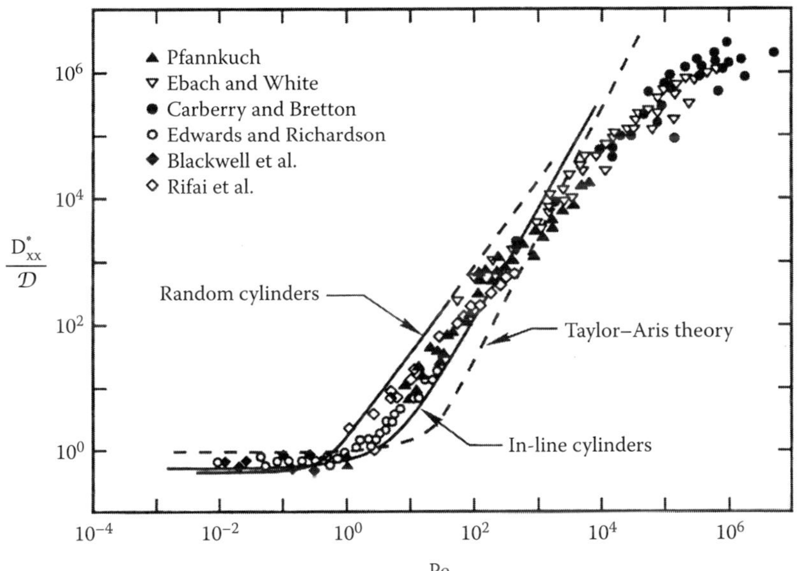

FIGURE 8.8 Experimental (points) and theoretical (lines) longitudinal dispersivities for packed beds. Most of the experiments are made with beds packed with monodisperse spheres. (From Whitaker S, *The Method of Volume Averaging*, Kluwer Academic Publishers, Dordrecht, the Netherlands, 1999, and references therein.)

and one with a specific concentration (c_0), as shown in Figure 8.9. In this case, a concentration profile in the y direction is induced by feeding liquid at two different concentrations over two sections of the inlet. The concentration profile in the y direction is measured at various x positions and at steady state.

In this case, there is symmetry in the z direction and Equation 8.37 simplifies to

$$u\frac{\partial c}{\partial x} = D_L^*\frac{\partial^2 c}{\partial x^2} + D_T^*\frac{\partial^2 c}{\partial y^2} \tag{8.52}$$

Furthermore, if the measuring probes are close enough to the inlet, we can invoke a boundary-layer type of approximation and neglect longitudinal dispersive transport:

$$D_L^*\frac{\partial^2 c}{\partial x^2} \ll D_T^*\frac{\partial^2 c}{\partial y^2} \tag{8.53}$$

So that Equation 8.52 simplifies to

$$u\frac{\partial c}{\partial x} = D_T^*\frac{\partial^2 c}{\partial y^2} \tag{8.54}$$

Appropriate boundary conditions are

$$c = \begin{cases} c_0, \ y \geq 0 \\ 0, \ y < 0 \end{cases}, \ x = 0 \tag{8.55}$$

$$c = c_0, \ y \to \infty \tag{8.56}$$

$$c = 0, \ y \to -\infty \tag{8.57}$$

where it has been assumed that the bed is wide enough in the y direction at the sampling points so that conditions 8.57 and 8.58 apply.

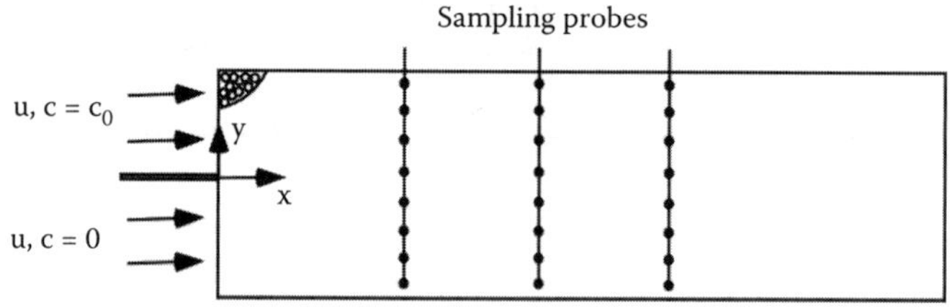

FIGURE 8.9 Experiment to determine the lateral dispersivity in a packed bed: a lateral concentration gradient is induced at the entrance. The concentration of dispersing species is measured as a function of x and y at sampling points.

The solution of the problem is

$$\frac{c}{c_0} = 1 - \frac{1}{2}\operatorname{erfc}\left(\frac{y}{2}\sqrt{\frac{u}{D_T^* x}}\right) \tag{8.58}$$

To put in dimensionless form, define

$$X = \frac{ux}{D_T^*} \tag{8.59}$$

$$Y = \frac{uy}{D_T^*} \tag{8.60}$$

Equation 8.58 becomes

$$\frac{c}{c_0} = 1 - \frac{1}{2}\operatorname{erfc}\left(\frac{Y}{2\sqrt{X}}\right) \tag{8.61}$$

Figure 8.10 shows a plot of these profiles.

Figure 8.11 shows experimentally determined lateral dispersivities for beds of spheres. Note that values are appreciably lower than longitudinal dispersivities (Figure 8.8) at the same Pe_p.

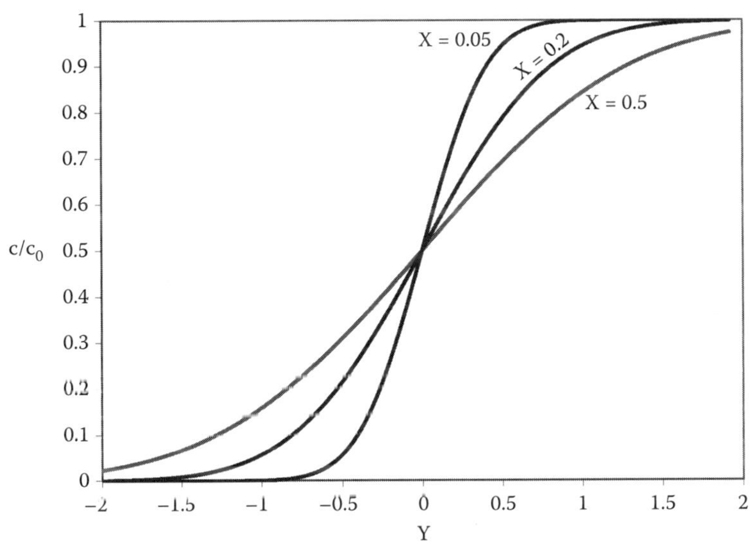

FIGURE 8.10 Dimensionless lateral concentration profiles at various dimensionless axial positions.

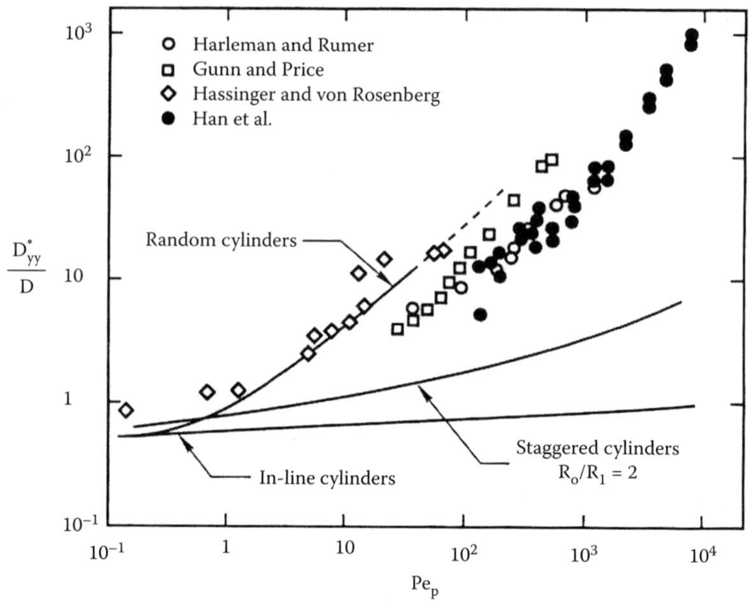

FIGURE 8.11 Experimental (points) and theoretical (lines) lateral dispersivities for packed beds. Most of the experiments are made with beds packed with monodisperse spheres. (From Whitaker S, *The Method of Volume Averaging*, Kluwer Academic Publishers, Dordrecht, the Netherlands, 1999, and references therein.)

8.5 TRANSPORT IN POROUS MEDIA: FLUID/SOLID INTERACTIONS

In the previous section we treated the case of passive dispersion, defined by the fact that the species whose transport is described does not participate in chemical reactions or interact with the solid phase. In most applications, interactions with the solid phase are important and the reason why a porous medium is used in engineered systems. In this section we study the transport of a species that interacts with the solid phase. The main interactions of interest occur when the species binds to the solid surface through physical (e.g., electrostatic) or chemical (e.g., covalent bonds) interactions or when the species undergoes a chemical reaction on the surface of the solid. In the latter case, the solid may be a reactant or a catalyst.

Examples of applications in environmental processes are

1. Adsorption/desorption in packed beds. Adsorption is one of the most commonly used methods for water and air treatment: the fluid is passed through a packed bed filled with solid particles on which a contaminant can adsorb so that it gets separated from the fluid. Desorption is the opposite process: a species adsorbed on the solid surface is transferred to the fluid. It is sometimes used to regenerate beds on which a contaminant is sorbed. Adsorption/desorption processes occur naturally in soils. Example

of widely used adsorption processes are: separation of volatile organic contaminants (VOCs) from water or air using activated carbon beds, and separation of arsenic from drinking water using iron hydroxide beds.

2. Ion exchange. This is a process similar to adsorption, but in this case the species to be adsorbed is an ion that replaces a different ion in the solid phase. This exchanged ion is transferred to the liquid.
3. Chromatography. Essentially, this is an adsorption/desorption process in which two or more species are adsorbed on the solid phase in a transient process designed to separate the various species by taking advantage of their different affinities for the solid phase: species that adsorb more strongly will be retained for a longer time in the bed.
4. Catalytic packed-bed reactors. Many chemical reactions need a catalyst to proceed to acceptable conversions. It is common to encounter catalysts deposited on solid surfaces that can accelerate chemical reactions whose reactants are in the fluid phase. In this type of process, the reactants typically adsorb on the surface first, then react, and finally the products desorb into the fluid phase. An example of this application is the conversion of halogenated hydrocarbons into alkanes using platinum as a catalyst in the presence of hydrogen gas (reductive dehalogenation).

To study the type of problems discussed above, we need to be able to formulate the transport equation for a chemical species that interacts with the solid. Consider the transport equation for the case of passive dispersion with one-dimensional flow (Equation 8.37). Let us take the special case in which c = c(t,x) (i.e., c is independent of y and z). The transport equation reduces to

$$\frac{\partial c}{\partial t} + u\frac{\partial c}{\partial x} = D_L^* \frac{\partial^2 c}{\partial x^2} \tag{8.62}$$

This equation can be derived by performing a molar shell balance on the transported species for a section of porous medium (Figure 8.12), as follows: let N* be the flux of species in the x direction, expressed as moles of species per unit time and per unit **area of porous medium**. Note that we have selected to define this flux using the whole area of medium as a basis, which includes both fluid and solid phases.

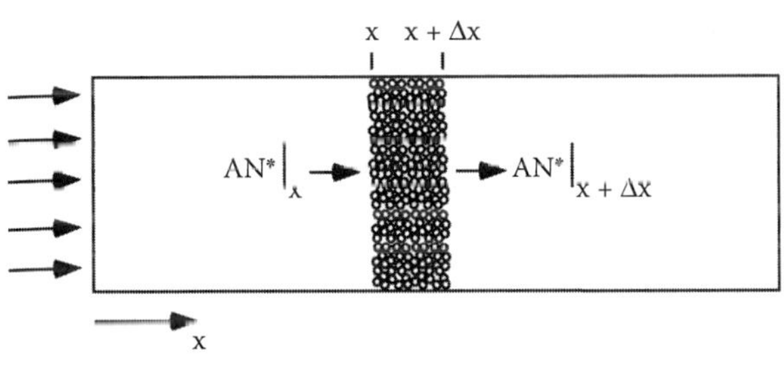

FIGURE 8.12 Molar shell balance in a porous medium.

For the case of passive dispersion, we can write a mole balance of species in the control volume shown in Figure 8.12 as follows:

$$\frac{\partial(cV_f)}{\partial t} = AN^*\Big|_x - AN^*\Big|_{x+\Delta x} \tag{8.63}$$

(accumulation) = (inlet) – (outlet)

The molar flux is interpreted as the superposition of convective and dispersive contributions. If we characterize these by means of the interstitial velocity and the dispersivity used before, the flux obtained would be per unit area of fluid. Therefore, we multiply each contribution by the porosity to obtain the flux,

$$N^* = \phi u - \phi D_L^* \frac{\partial c}{\partial x} \tag{8.64}$$

Also, we have

$$V_f = \phi V = \phi A\Delta x \tag{8.65}$$

Substituting Equations 8.64 and 8.65 into Equation 8.63, dividing by $\phi A\Delta x$, and taking the limit as $\Delta x \rightarrow 0$ recovers Equation 8.62.

This way of formulating the governing equation is useful to obtain the averaged continuity equation for cases in which there is fluid/solid transport of the species considered. In the following sections, we will treat these types of problems.

8.5.1 Equilibrium Adsorption in Porous Media: Linear Equilibrium

In this section we will consider that the species analyzed can adsorb on the surface of the solid phase. In addition, we will consider valid the following assumptions/conditions:

1. Adsorption is reversible and instantaneous, so that we can assume that the fluid in the immediate vicinity of solid surface is at equilibrium with the solid surface. Any internal porosity inside the particles that is accessible to the fluid does not produce diffusion limitations.
2. The adsorption isotherm is linear.
3. There are no mass transfer limitations in the fluid phase; that is the concentration of the fluid in the immediate vicinity of the solid phase is the bulk fluid concentration, c.

These conditions imply that, for a point in the porous medium continuum, the concentration on the surface of the solid phase can be expressed as

$$\hat{c}_s = K_{eq} c \tag{8.66}$$

where $\hat{c}_s$ is moles of species adsorbed per unit surface area of solid phase. If concentrations are expressed in mass rather than moles (c would be the intrinsic mass average concentration in the fluid: mass of species per unit fluid volume), then $\hat{c}_s$ is mass of species adsorbed per unit surface area of solid phase. It is customary to represent the concentration of adsorbed species in the solid phase as mass of species per unit mass of solid (q). Note that

$$q = \frac{\hat{c}_s a_s}{\rho_s} \tag{8.67}$$

where a_s is the specific surface area of solid phase, expressed as surface area of solid per unit volume of solid, and ρ_s is the density of the solid phase. Equations 8.66 and 8.68 imply that q is linearly related to c also.

The mole (or mass balance) for the transported species in the fluid phase of the control volume represented in Figure 8.12 is, in this case,

$$\frac{\partial(cV_f)}{\partial t} = AN^*\Big|_x - AN^*\Big|_{x+\Delta x} - J_s A_s \tag{8.68}$$

where J_s is the adsorption flux (moles or mass of species adsorbed per unit time and unit surface area of solid phase), and A_s is the surface area of solid phase in the control volume. Note that J_s is positive whenever the treated species is transferred from the fluid to the solid and negative in the opposite case.

Equation 8.68 is complemented by a mole (or mass) balance in the solid phase within the control volume:

$$\frac{\partial(\hat{c}_s A_s)}{\partial t} = J_s A_s \tag{8.69}$$

(note that we are not allowing for direct transport of species from solid particle to solid particle).

Now, we substitute Equations 8.64 and 8.65 into Equation 8.68, divide by $\varphi A\Delta x$, and take the limit as $\Delta x \rightarrow 0$ to obtain

$$\frac{\partial c}{\partial t} + u\frac{\partial c}{\partial x} = D_L^* \frac{\partial^2 c}{\partial x^2} - \frac{A_s}{\phi A\Delta x} J_s \tag{8.70}$$

Note that

$$A_s = a_s V_s = a_s(1-\phi)A\Delta x \tag{8.71}$$

so that Equation 8.71 can be expressed as

$$\frac{\partial c}{\partial t} + u\frac{\partial c}{\partial x} = D_L^* \frac{\partial^2 c}{\partial x^2} - \frac{(1-\phi)a_s}{\phi} J_s \tag{8.72}$$

Let

$$a_v = \frac{(1-\phi)a_s}{\phi} \tag{8.73}$$

Note that a_v represents the surface area of solid per unit volume of fluid phase. Equation 8.72 can be expressed as

$$\frac{\partial c}{\partial t} + u\frac{\partial c}{\partial x} = D_L^* \frac{\partial^2 c}{\partial x^2} - a_v J_s \tag{8.74}$$

On the other hand, Equation 8.69 can be written as

$$\frac{\partial \hat{c}_s}{\partial t} = J_s \tag{8.75}$$

Equations 8.74 and 8.75 are the governing equations for transport with adsorption in porous media. In the particular case of interest, adsorption is considered to be so fast that equilibrium is instantaneously established between the fluid and the solid phases and Equation 8.66 applies. Substituting Equation 8.66 into Equation 8.75 leads to

$$K_{eq}\frac{\partial c}{\partial t} = J_s \tag{8.76}$$

This expression for the adsorption flux can be substituted into Equation 8.74. After manipulations, the resulting equation can be written as

$$\frac{\partial c}{\partial t} + \frac{u}{R}\frac{\partial c}{\partial x} = \frac{D_L^*}{R}\frac{\partial^2 c}{\partial x^2} \tag{8.77}$$

where

$$R = 1 + K_{eq} a_v \tag{8.78}$$

Note that Equation 8.77 is mathematically identical to the equation governing passive dispersion (Equation 8.43). The only difference is that the coefficients (u and D_L^*) appear divided by the parameter $R > 1$. Therefore, species transport in this case resembles passive dispersion, but with a lower velocity and dispersivity. For this reason, the parameter R is called **retardation factor**. The mathematical similarity allows us to apply solutions of Equation 8.43 to this process. For example, for the case of a long bed initially containing no contaminant to which a concentration c_0 is fed after $t = 0$, the conditions given by Equations 8.44 through 8.46 apply, and the solution is obtained by modifying Equation 8.47; that is,

$$\frac{c}{c_0} = \frac{1}{2}\text{erfc}\left(\frac{x - ut/R}{2\sqrt{D_L^* t/R}}\right) + \frac{1}{2}\exp\left(\frac{ux}{D_L^*}\right)\text{erfc}\left(\frac{x + ut/R}{2\sqrt{D_L^* t/R}}\right) \tag{8.79}$$

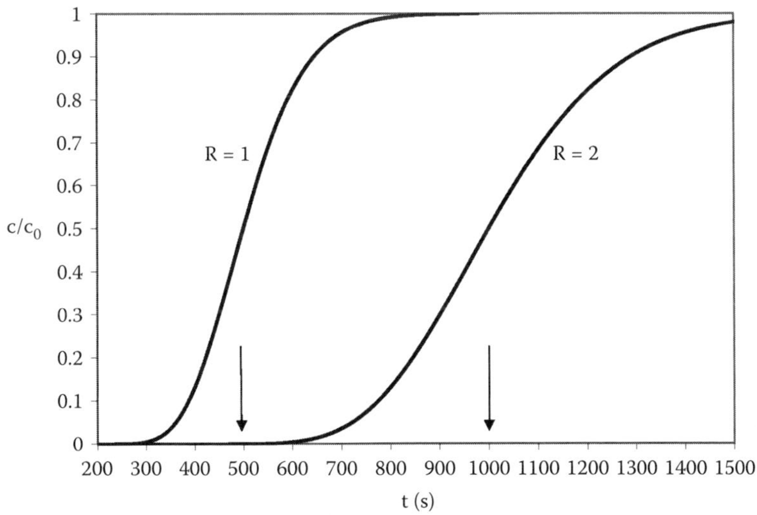

FIGURE 8.13 Breakthrough curves for the example indicated in the text for passive dispersion (R = 1) and adsorption (R = 2). The arrows represent residence times of the species in the bed.

As an example, consider a bed packed with monodisperse size spheres with $d = 0.1$ cm and a porosity $\pi = 0.4$, through which a liquid is flowing at $u = 0.1$ cm/s. The diffusivity of the contaminant in the liquid is $D = 1 \times 10^{-5}$ cm^2/s and we would like to determine the concentration as a function of time at $x = 50$ cm for: (a) passive dispersion, (b) equilibrium linear adsorption with $R = 2$. First, we estimate the longitudinal dispersivity. To do this, we calculate the Peclet number by Equation 8.32:

$$Pe_p = \frac{ud\pi}{D(1-\pi)} = \frac{0.1 \times 0.1 \times 0.4}{10^{-5} \times 0.6} = 6667 \tag{8.80}$$

From Figure 8.8 we find

$$\frac{D_L^*}{D} \approx 10^4 \Rightarrow D_L^* = 0.1\text{cm}^2/\text{s} \tag{8.81}$$

Now, we evaluate the concentration as a function of time at $x = 50$ cm from Equation 8.79. The results are presented in Figure 8.13.

8.5.2 Equilibrium Adsorption in Porous Media: Nonlinear Equilibrium

The analysis presented above can be generalized to any adsorption isotherm as follows. First, Equation 8.66 is replaced by the more general expression

$$\hat{c}_s = F(c) \tag{8.82}$$

where F represents the adsorption isotherm. Now, we take the time derivative of this equation and employ the chain rule to get

$$\frac{\partial \hat{c}_s}{\partial t} = \frac{dF}{dc}\frac{\partial c}{\partial t} = F'(c)\frac{\partial c}{\partial t} \tag{8.83}$$

Substituting this equation into Equation 8.75 leads to

$$F'(c)\frac{\partial c}{\partial t} = J_s \tag{8.84}$$

Now, we substitute this equation into Equation 8.62 and rearrange to get

$$\frac{\partial c}{\partial t} + \frac{u}{[1 + a_v F'(c)]}\frac{\partial c}{\partial x} = \frac{D_L^*}{[1 + a_v F'(c)]}\frac{\partial^2 c}{\partial x^2} \tag{8.85}$$

which can be interpreted as a process with a retardation factor that depends on concentration. The nonlinear nature of the differential Equation 8.85 implies that a numerical method is required for its solution.

8.5.3 Transport in Porous Media with Rate-Limited Adsorption and/or Chemical Reaction

Consider a process in which the transported species adsorbs to the solid phase, but now the adsorption is slow, so that its rate needs to be considered in the analysis. Physically, this means that the liquid and solid phases do not achieve equilibrium. In this case, the balances represented by Equations 8.74 and 8.75 still apply, and a constitutive equation is needed to represent the adsorption flux, J_s; that is, the kinetics of the adsorption process is needed. We still are assuming that there are no mass transfer limitations in the fluid phase. Consider the case in which adsorption is reversible and the kinetic expression is first order for both adsorption and desorption rates:

$$J_s = k_a c - k_d \hat{c}_s \tag{8.86}$$

where k_a and k_s are adsorption and desorption rate constants, respectively. Equations 8.74 and 8.75 become

$$\frac{\partial c}{\partial t} + u\frac{\partial c}{\partial x} = D_L^* \frac{\partial^2 c}{\partial x^2} - a_v(k_a c - k_d \hat{c}_s) \tag{8.87}$$

$$\frac{\partial \hat{c}_s}{\partial t} = k_a c - k_d \hat{c}_s \tag{8.88}$$

These two differential equations would have to be solved numerically to find simultaneously c(t,x) and $\hat{c}_s$(t,x).

A particular case occurs when the rate of adsorption is much faster than the rate of desorption. This could be a consequence of: (i) the partitioning of the transferred species favors the solid, (ii) the adsorption process is far from equilibrium (the problem is being analyzed at a time scale for which $\hat{c}_s$ is so small that the desorption rate is negligible), or (iii) adsorption is irreversible. In this case we can neglect the rate of desorption in Equation 8.87 and the governing equation for the fluid concentration becomes decoupled from the equation for the solid concentration; that is,

$$\frac{\partial c}{\partial t} + u\frac{\partial c}{\partial x} = D_L^* \frac{\partial^2 c}{\partial x^2} - a_v k_a c \tag{8.89}$$

Note that this is the same equation that would be obtained for the case in which a first-order chemical reaction involving the disappearance of the transported species occurred on the surface of the solid phase (catalytic reaction). Also, if the transported species were subjected to a first-order, homogeneous chemical reaction in the liquid phase, a similar equation would be obtained, replacing $a_v k_a$ by the kinetic rate constant of the homogeneous reaction.

Equation 8.89 subject to the initial and boundary conditions given by Equations 8.44 through 8.46 has the following analytical solution:

$$\frac{c}{c_0} = \frac{1}{2}\exp\left[\frac{x(u-\alpha)}{2D_L^*}\right]\left[\text{erfc}\left(\frac{x-\alpha t}{2\sqrt{D_L^* t}}\right) + \exp\left(\frac{\alpha x}{D_L^*}\right)\text{erfc}\left(\frac{x+\alpha t}{2\sqrt{D_L^* t}}\right)\right] \tag{8.90}$$

where

$$\alpha = \sqrt{u^2 + 4a_v k_a D_L^*} \tag{8.91}$$

Note that when $t \to \infty$ the following steady-state concentration profile is obtained (recall that erfc(∞) = 0, erfc(−∞) = 2)

$$c = c_0 \exp\left[\frac{x(u-\alpha)}{2D_L^*}\right], \text{ steady-state} \tag{8.92}$$

The steady-state concentration is a monotonically decreasing function of x (note that $\alpha > u$).

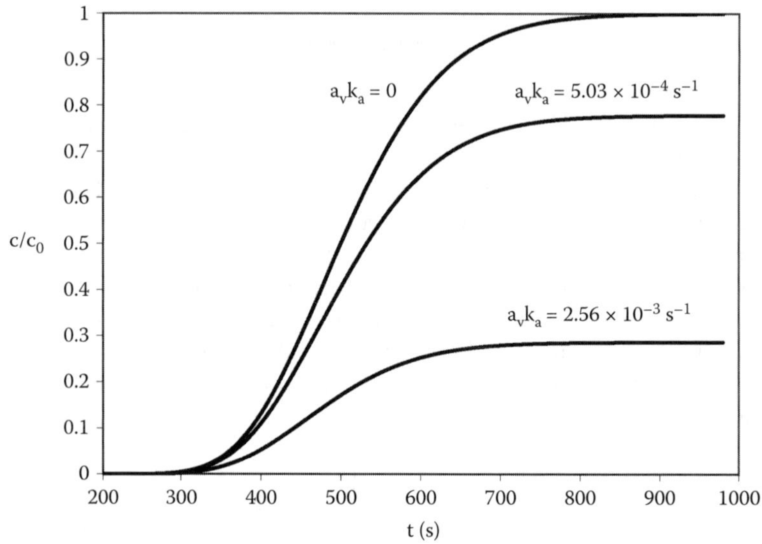

FIGURE 8.14 Breakthrough curves for the example considered in Figure 8.13 with and without irreversible adsorption/reaction at the indicated rates.

Figure 8.14 shows plots of dimensionless concentration at $x = 50$ cm for the example treated before (Figure 8.13), and various values of the reaction/adsorption parameter. The steady-state concentration decreases as the reaction rate increases, as expected.

8.5.4 Adsorption and Reaction in Porous Media with Mass Transfer Limitations

All the cases analyzed so far about fluid/solid interactions have assumed that the surface of the solid is exposed to liquid at its bulk (interstitial) concentration. If the processes occurring at the fluid/solid interface are fast enough, a concentration gradient in the fluid phase may be generated at the pore level. Essentially, this would mean that mixing in the fluid at the pore level is not perfect, and local mass transfer limitations will play a role in the description of transport.

There are two important types of mass transfer limitations: convective mass transfer in the interstitial space transporting species from the bulk fluid to the solid surface, and, for the case of particles with internal porosity, diffusion of species from the surface of the particles to the internal pore space. We will consider both cases separately.

8.5.4.1 Mass Transfer Limitations: Convection at the Pore Level

The general mole (mass) balance for the fluid phase is still represented by Equation 8.74 for one-dimensional flow in a porous medium. Mass transfer limitations will be reflected in the choice of equation to represent the flux J_s. Here, we will use a convective mass transfer coefficient to represent the flux. Let c_s be the concentration of transferred species in the fluid at the immediate vicinity of the solid phase. We can express the flux as

$$J_s = k_m(c - c_s) \tag{8.93}$$

where k_m is the convective mass transfer coefficient. Equation 8.74 can be written as

$$\frac{\partial c}{\partial t} + u\frac{\partial c}{\partial x} = D_L^* \frac{\partial^2 c}{\partial x^2} - a_v k_m (c - c_s) \tag{8.94}$$

In addition, we need a governing equation for c_s. Here, we will assume that the fluid adjacent to the solid phase reaches instantaneous equilibrium with the solid surface. We will consider the case of a linear adsorption isotherm, so that

$$\hat{c}_s = K_{eq} c_s \tag{8.95}$$

A balance of species in the solid phase is given by Equation 8.75, which, after combining it with Equation 8.95, becomes

$$\frac{\partial \hat{c}_s}{\partial t} = k_m(c - c_s) \tag{8.96}$$

Using Equation 8.95, Equation 8.96 can be written as follows:

$$\frac{\partial c_s}{\partial t} = \frac{k_m}{K_{eq}}(c - c_s) \tag{8.97}$$

Equations 8.94 and 8.95, along with appropriate initial and boundary conditions, represent the formulation of the problem, which would have to be solved numerically. Note that this formulation is mathematically similar to the problem of nonequilibrium reversible adsorption treated before (Equations 8.87 and 8.88).

If adsorption is irreversible and fast (compared with the mass transfer rate), or if the transferred species is consumed by a chemical reaction that occurs on the surface of the solid, and the reaction is fast, we will have $c_s = 0$ and Equation 8.94 simplifies to

$$\frac{\partial c}{\partial t} + u\frac{\partial c}{\partial x} = D_L^* \frac{\partial^2 c}{\partial x^2} - a_v k_m c \tag{8.98}$$

This equation is mathematically identical to Equation 8.89, so that the solution given by Equation 8.90 applies, replacing k_a by k_m. The mathematical analogy brings up an important question. Suppose that the chemical reaction that occurs on the surface of the solid is first order. If the reaction kinetics and mass transfer coefficient are unknown, how can we decide whether the process is mass transfer limited or reaction limited? Answer: Equation 8.90 can be fitted to the data regardless of which process occurs. If the process is reaction limited, the fitting constant (k_a in Equation 8.89) would be the first-order kinetic rate constant, and if the process is mass-transfer controlled, then the fitting constant would be the mass transfer coefficients. The value of k_m must depend on fluid velocity; in fact, one would expect mass transfer limitations to eventually disappear (i.e., $k_m \rightarrow \infty$) if the fluid velocity is high enough. One could carry out a series of experiments for various fluid velocities and find the fitting constant. If its value does not change with fluid velocity, then the fitting constant would represent a kinetic rate. Otherwise, it would represent a mass transfer coefficient.

8.5.4.2 Mass Transfer Limitations: Intraparticle Diffusion

A large number of porous media transport processes dealing with adsorption or catalytic reactions are carried out using porous solid particles, since the internal porosity substantially increases the surface area of solid phase. In these cases, transport inside the particle occurs by diffusion only and this is typically a rate-limiting step in the process. A convenient way to deal with intraparticle diffusion is through the use of the effectiveness factor. In general, the definition of effectiveness factor for a solid particle surrounded by a fluid is

$$\eta = \frac{\left\{\begin{array}{l}\text{Rate of consumption of species} \\ \text{(moles per unit time) in the particle}\end{array}\right\}}{\left\{\begin{array}{l}\text{Rate of consumption of species that would be} \\ \text{obtained if the concentration of species in the particle} \\ \text{were uniform and equal to the surface concentration}\end{array}\right\}} \tag{8.99}$$

For an irreversible first-order reaction with no external mass transfer limitations, this definition implies that

$$J_s = \eta k c \tag{8.100}$$

where k is the heterogeneous kinetic rate constant. Equation 8.74 then becomes

$$\frac{\partial c}{\partial t} + u\frac{\partial c}{\partial x} = D_L^* \frac{\partial^2 c}{\partial x^2} - a_v \eta k c \tag{8.101}$$

Once again, this equation is mathematically identical to Equation 8.89, provided that the effectiveness factor is independent of concentration. For first-order reactions it can be shown that (see the analysis of transport in a biofilm in Chapter 6)

$$\eta = \eta(\text{particle geometry}, \pi_T) \tag{8.102}$$

where π_T is the Thiele modulus, defined as

$$\pi_T = \ell\sqrt{\frac{k}{D_{eff}}} \tag{8.103}$$

Here, ℓ is a characteristic length of the catalyst particle and D_{eff} is the effective diffusivity of the transported species in the catalyst particle.

The governing Equation 8.101 is identical to Equations 8.89 and 8.98, but the physical meaning of the constant in the sink term is different. An experimental test to corroborate if internal diffusion limitations control the process would involve carrying out experiments in which the effectiveness factor is changed. For example, this could be accomplished by changing particle size (ℓ).

PROBLEMS

8.1 Researchers have developed a new iron hydroxide-based sorbent to remove arsenic from drinking water. To demonstrate the technology, they set up a packed-bed adsorption experiment. The sorbent consists of non-porous, spherical solid particles with an average diameter of 2 mm. The bed porosity is 0.4. The length of the bed is 50 cm. The species to be adsorbed is arsenate ion, which has a diffusivity in water approximately equal to 10^{-5} cm²/s. The superficial velocity used in the experiment is 1 cm/s. The experiment consists in continuously feeding the bed with a sample of water with a particularly high arsenic content: 100 μg/L (the EPA standard for arsenic in drinking water is 10 μg/L). When the experiment is finished, the developers of the sorbent make the following claims, based on experimental observations and data obtained:

Claim 1: There were no mass transfer limitations in the experiment, and adsorption is governed by linear equilibrium.

Claim 2: Breakthrough of arsenic from the bed occurred at 17,000 bed volumes.

Claim 3: The experiment was terminated at 23,000 bed volumes, moment at which the exit concentration was 10 μg/L.

The number of bed volumes (B) is defined as the ratio between the total volume of solution that has passed through the bed from t = 0 to a given time and the total bed volume; that is,

$$B = \frac{Qt}{V}$$

where Q is volumetric flow rate.

The breakthrough point is defined as the time for which the exit concentration reaches 1% of the feed concentration.

Use Figure 8.8 to estimate the longitudinal effective diffusivity.

a. From assumptions based on claim 1 and the information in claim 2, estimate the partition coefficient for arsenate between water and the solid (K_{eq}).

b. Can you corroborate claim 3?

8.2 Our objective is to simulate the separation of two solutes A and B in a liquid chromatography column. The column is a bed ($\pi = 0.5$) packed with spheres ($d = 1$ mm) and a total length $L = 2$ m. A carrier liquid flows continuously through the column. At $t = 0$ a slug of liquid containing the same number of moles of A and B ($n = 1$ mol of each) is injected into the carrier fluid at the entrance of the column ($x = 0$). The carrier fluid flows through the column at an intrinsic phase average velocity $u = 1$ cm/s. The diffusivity of both solutes is the same: $D_A = D_B = 10^{-5}$ cm²/s. The cross-sectional area of column open to the fluid is $A_f = 0.1$ cm².

Both solutes experience equilibrium adsorption at the solid surface. The solution of the convective–dispersion equation for the inlet condition considered here is

$$c = \frac{nxR}{2uA_f\sqrt{\pi D_L^* t^3/R}} \exp\left[-\frac{R}{4D_L^* t}\left(x - \frac{ut}{R}\right)^2\right]$$

The retardation factors are: $R_A = 3$, $R_B = 8$.

a. Plot c_A and c_B vs. t (in the same plot) at the column exit.

b. The peak elution time (t_e) is the time at which the solute concentration at the exit reaches a maximum as a function of time. Find an expression to calculate the peak elution time for a solute (check your result with the plots in part 1).

References

Adler PM, *Porous Media. Geometry and Transports*, Butterworth-Heinemann, Boston, 1992.

Bird RB, Stewart WE, Lightfoot EN, *Transport Phenomena*, John Wiley & Sons, New York, 2nd ed, 2007.

Brown R, A brief account of microscopical observations made on the particles contained in the pollen of plants, *Philos Mag*, 4, 161–173, 1828.

Chen CL, JH Wu, IC Tseng, TM Liang, WT Liu, Characterization of active microbes in a full-scale anaerobic fluidized bed reactor treating phenolic wastewater, *Microbes Environ*, 24, 144–153, 2009.

Clark MM, *Transport Modeling for Environmental Engineers and Scientists*, John Wiley & Sons, New York, 1996.

Cussler EL, *Diffusion. Mass Transfer in Fluid Systems*, Cambridge University Press, Cambridge, 2nd ed, 1997.

Das S, Bhantia AK, Adhikari B, Removal of chlorinated volatile organic contaminants from water by pervaporation using a novel polyurethane urea–poly (methyl methacrylate) interpenetrating network membrane, *Chem Eng Sci*, 61, 6454–6467, 2006.

Einstein A, *Investigations on the Theory of the Brownian Movement*, edited with notes by R. Fürth, Dover Publications, New York, 1956.

EPA, *Cross-flow Pervaporation Technology*, EPA/540/R-95/511, 1998.

Felder RN, Rousseau RW, *Elementary Principles of Chemical Processes*, John Wiley & Sons, New York, 2005.

Fick A, On liquid diffusion, *Poggendorff's Annalen*, 94, 59–86, 1855a, reprinted in *J Membr Sci*, 100, 33–38, 1995.

Fick A, On liquid diffusion, *Philos Mag*, 10, 30–39, 1855b, reprinted in *J Membr Sci*, 100, 33–38, 1995.

Geankoplis CJ, *Transport Processes and Separation Process Principles*, Prentice Hall, Englewood Cliffs, New Jersey, 1993.

Gupta SC, *The Classical Stefan Problem*, Elsevier, Amsterdam, 2003.

Kim KS, Kwon TS, Yang JS, Simultaneous removal of chlorinated contaminants by pervaporation for the reuse of a surfactant, *Desalination*, 205, 87–96, 2007.

Kreyszig E, *Advanced Engineering Mathematics*, John Wiley & Sons, New York, 1999.

Philibert J, One and a half century of diffusion: Fick, Einstein, before and beyond, *Diffusion Fundam*, 4, 6.1–6.19, 2006.

Reid RC, Pransuitz JM, and Poling BE, *The Properties of Gases and Liquids*, McGraw-Hill, New York, 1987.

Rubinstein LI, *The Stefan Problem, Translations of Mathematical Monographs 27*, American Mathematical Society, Providence, RI, 1971.

Smoluchowski MM, Essai d'une théorie cinétique du muovement Brownien et des milieux troubles, *Bull Int Acad Sci Cracovie*, 577–602, 1906.

Whitacre DM, *Reviews of Environmental Contamination and Toxicology*, Springer, New York, 2013.

Whitaker S, *The Method of Volume Averaging*, Kluwer, Dordrecht, the Netherlands, 1999.

White FM, *Fluid Mechanics*, McGraw-Hill, New York, 2nd ed, 1986.

Appendix A: Vectors and Vector Operations

Here, we are mostly concerned with **intensive or point** physical quantities that represent a physical property of a material and that can be defined at each point in space. Physical quantities that are characterized by numbers are called scalars. For example, the temperature of a material is completely known by specifying its value in a specific temperature scale. Physical quantities that require a value and a direction in space are called vectors. We will denote vectors by bold-face symbols. For example, the velocity of a material at a point is specified by the velocity vector, **v**. Vectors can be represented in terms of their components with respect to a coordinate system. For example, a vector (Figure A.1) can be expressed as follows.

$$\mathbf{a} = a_x \mathbf{e}_x + a_y \mathbf{e}_y + a_z \mathbf{e}_z \tag{A.1}$$

where a_x, a_y, and a_z are the components of the vector, and $\mathbf{e}_x$, $\mathbf{e}_y$, and $\mathbf{e}_z$ are unit vectors that represent the three independent directions in space x, y, and z, as defined by the particular set of coordinates.

Alternatively, a vector can be specified by giving its magnitude,

$$a = |\mathbf{a}| = \sqrt{a_x^2 + a_y^2 + a_z^2} \tag{A.2}$$

and its direction in space. Vectors with unit magnitude are called **unit vectors**.

Important vector operations are as follows.

A.1 SCALAR (DOT) PRODUCT BETWEEN TWO VECTORS

The scalar product between two vectors can be represented mathematically in two equivalent ways. First, it is equal to the sum of the products of the vectors' components:

$$\mathbf{a} \cdot \mathbf{b} = a_x b_x + a_y b_y + a_z b_z \tag{A.3}$$

The scalar product also can be expressed as

$$a \cdot b = ab \cos \theta \tag{A.4}$$

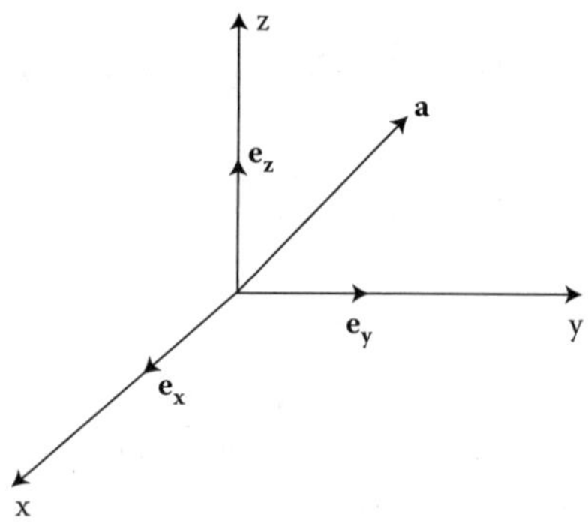

FIGURE A.1 A vector (**a**) in a rectangular coordinate system.

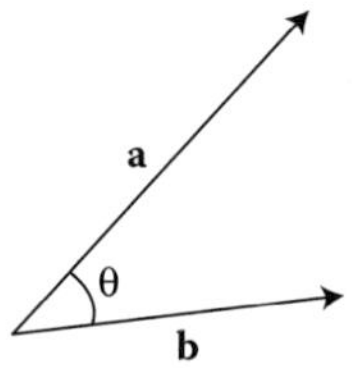

FIGURE A.2 Two vectors in space form an angle θ between them.

where θ is the angle between the two vectors (Figure A.2). Note that the scalar product between a vector and itself (θ = 0) is the square of its magnitude. Note also that the scalar product between two perpendicular vectors is zero.

By construction, it should be realized that the scalar product represents the projection of one vector onto the direction specified by the other.

A.2 VECTOR (CROSS) PRODUCT BETWEEN TWO VECTORS

In terms of components, the cross product between two vectors can be expressed in terms of a determinant,

$$\mathbf{a} \times \mathbf{b} = \begin{vmatrix} \mathbf{e}_x & \mathbf{e}_y & \mathbf{e}_z \\ a_x & a_y & a_z \\ b_x & b_y & b_z \end{vmatrix} \tag{A.5}$$

In expanded form, the determinant is

$$\mathbf{a} \times \mathbf{b} = \mathbf{e}_x(a_y b_z - b_y a_z) - \mathbf{e}_y(a_x b_z - b_x a_z) + \mathbf{e}_z(a_x b_y - b_x a_y) \tag{A.6}$$

Note that the vector product between two vectors is another vector. Alternatively, the magnitude of the cross product is

$$\left|\mathbf{a} \times \mathbf{b}\right| = ab\sin\theta \tag{A.7}$$

and the direction of a × b is perpendicular to both **a** and **b** (i.e., perpendicular to the plane formed by the vectors **a** and **b**) with direction given by the right-hand rule. Note that the cross product between a vector and itself is zero.

A.3 DIFFERENTIAL OPERATIONS

Physical quantities (scalar or vector) can be a function of position and time. For example, a scalar quantity f can be expressed as f = f(t,x,y,z) in a rectangular coordinate system. Spatial variations of f can be quantified by the **gradient** of f, defined by

$$\nabla f = \frac{\partial f}{\partial x}\mathbf{e}_x + \frac{\partial f}{\partial y}\mathbf{e}_y + \frac{\partial f}{\partial z}\mathbf{e}_z \tag{A.8}$$

The gradient of the scalar quantity can be used to determine the rate of change of the quantity in a given direction in space. To specify a direction, we can use a unit vector **n** pointing in the direction in question. The rate of change of the quantity is called **directional derivative** and is calculated from

$$\frac{\partial f}{\partial n} = \mathbf{n} \cdot \nabla f = n_x \frac{\partial f}{\partial x} + n_y \frac{\partial f}{\partial y} + n_z \frac{\partial f}{\partial z} \tag{A.9}$$

For a vector field that is a function of position and time, **v**(t,x,y,z), the **divergence** of the vector field is defined by

$$\nabla \cdot \mathbf{v} = \frac{\partial v_x}{\partial x} + \frac{\partial v_y}{\partial y} + \frac{\partial v_z}{\partial z} \tag{A.10}$$

Note that the divergence is a scalar function of position and time.

The **curl** of a vector field is defined by

$$\nabla \times \mathbf{v} = \begin{vmatrix} \mathbf{e}_x & \mathbf{e}_y & \mathbf{e}_z \\ \partial/\partial x & \partial/\partial y & \partial/\partial z \\ v_x & v_y & v_z \end{vmatrix} \tag{A.11}$$

In expanded form this equation is

$$\nabla \times \mathbf{v} = \mathbf{e}_x\left(\frac{\partial v_z}{\partial y} - \frac{\partial v_y}{\partial z}\right) - \mathbf{e}_y\left(\frac{\partial v_z}{\partial x} - \frac{\partial v_x}{\partial z}\right) + \mathbf{e}_z\left(\frac{\partial v_y}{\partial x} - \frac{\partial v_x}{\partial y}\right) \tag{A.12}$$

Appendix B: The Continuity Equation

In Chapter 2, we used the principle of conservation of mass to derive the macroscopic mass balance. Here, we will look for the consequence of mass conservation at each point in a fluid. The final result will be an equation that is applicable at each point named the continuity equation. We start by defining the control volume shown in Figure B.1: a parallelepiped of sides $\Delta x \times \Delta y \times \Delta z$. Now, we write an instantaneous mass balance:

$$\left\{\begin{matrix}\text{Change of the total mass in the} \\ \text{control volume per unit time} \\ \text{(accumulation)}\end{matrix}\right\} = \{\text{Mass flow rate in}\} - \{\text{Mass flow rate out}\} \tag{B.1}$$

The time rate of change of the total mass can be expressed in terms of the fluid density as follows:

$$\left\{\begin{matrix}\text{The time rate of change of the} \\ \text{total mass in a control volume}\end{matrix}\right\} = \frac{dM}{dt} = \frac{d}{dt}(\rho\Delta x\Delta y\Delta z) = \Delta x\Delta y\Delta z\frac{\partial\rho}{\partial t} \tag{B.2}$$

where the use of partial derivative for the density reflects the fact that it may depend on position and time. To evaluate the mass flow rates in and out of the control volume, we have to consider the mass flow rates crossing each of the six surfaces that delimit the control volume: the mass flow rate through each surface is $\rho v_\perp \Delta A$, where $v_\perp$ is the component of the fluid velocity perpendicular to the surface (which represents the fluid crossing the surface) and ΔA the area of the surface. Each mass flow rate is represented in Figure B.2. Accordingly, the right-hand side of Equation B.1 can be expressed as

$$\left\{\begin{matrix}\text{The } \textbf{net} \text{ rate at which mass} \\ \textbf{enters} \text{ the control volume}\end{matrix}\right\} = \rho v_x\big|_x \Delta y\Delta z + \rho v_y\big|_y \Delta x\Delta z + \rho v_z\big|_z \Delta x\Delta y$$
$$-\rho v_x\big|_{x+\Delta x} \Delta y\Delta z - \rho v_y\big|_{y+\Delta y} \Delta x\Delta z - \rho v_z\big|_{z+\Delta z} \Delta x\Delta y \tag{B.3}$$

Substituting Equations B.2 and B.3 into Equation B.1, and dividing by $\Delta x\Delta y\Delta z$ yields

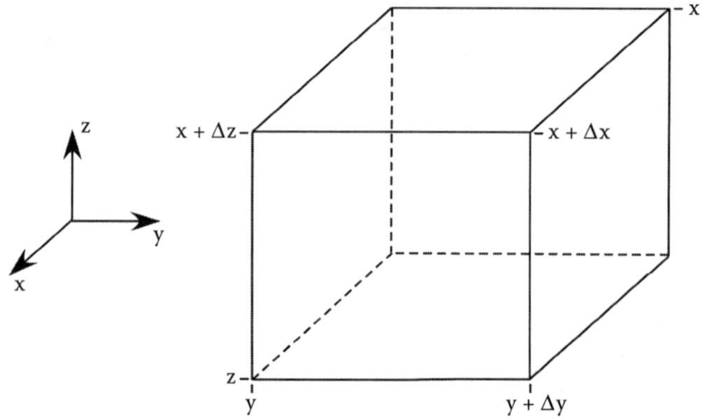

FIGURE B.1 Control volume of differential size used in the application of mass conservation.

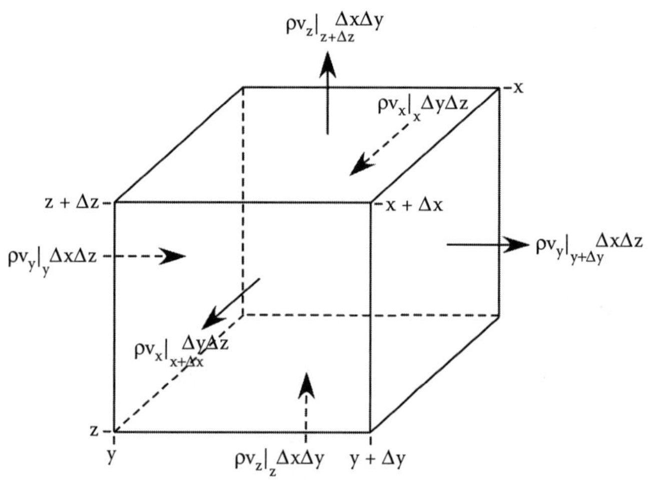

FIGURE B.2 Mass flow rates that cross each surface of the control volume.

$$\frac{\partial \rho}{\partial t} + \frac{\rho v_x|_{x+\Delta x} - \rho v_x|_x}{\Delta x} + \frac{\rho v_y|_{y+\Delta y} - \rho v_y|_y}{\Delta y} + \frac{\rho v_z|_{z+\Delta z} - \rho v_z|_z}{\Delta z} = 0 \quad \text{(B.4)}$$

Now, we take the limit as $\Delta x, \Delta y, \Delta z \to 0$ to obtain

$$\frac{\partial \rho}{\partial t} + \frac{\partial(\rho v_x)}{\partial x} + \frac{\partial(\rho v_y)}{\partial y} + \frac{\partial(\rho v_z)}{\partial z} = 0 \quad \text{(B.5)}$$

In vector notation, this equation can be written using the divergence operator (see Appendix A) as follows:

$$\frac{\partial \rho}{\partial t} + \nabla \cdot (\rho \mathbf{v}) = 0 \quad \text{(B.6)}$$

This is the **continuity equation**, which is the principle of mass conservation applied to a point in the fluid continuum.

In many applications, the density of the fluid can be considered constant and uniform. Most isothermal flows of liquids and some flows of gases with moderate pressure changes will follow this behavior. These flows, in which a given volume of fluid always contains the same mass, are called **incompressible flows**. In this case, the continuity equation simplifies as follows:

$$\frac{\partial \rho}{\partial t} = 0,\ \nabla \cdot (\rho \mathbf{v}) = \rho \nabla \cdot \mathbf{v} \Rightarrow\ \nabla \cdot \mathbf{v} = 0, \text{ incompressible flows} \quad \text{(B.7)}$$

In rectangular (Cartesian) coordinates, this equation reads

$$\frac{\partial v_x}{\partial x} + \frac{\partial v_y}{\partial y} + \frac{\partial v_z}{\partial z} = 0 \quad \text{(B.8)}$$

Appendix C: Mass Transfer Point Equations

C.1 BINARY SYSTEMS

Species A continuity equation:

$$\frac{\partial c_A}{\partial t} + \nabla \cdot \mathbf{N}_A = R_A \tag{C.1}$$

Molar flux equation:

$$\mathbf{N}_A = x_A(\mathbf{N}_A + \mathbf{N}_B) + \mathbf{J}_A^* \tag{C.2}$$

Fick's law:

$$\mathbf{J}_A^* = -cD_{AB}\nabla x_A \tag{C.3}$$

C.1.1 Species A Continuity Equation in Different Coordinate Systems

Rectangular:

$$\frac{\partial c_A}{\partial t} + \frac{\partial N_{Ax}}{\partial x} + \frac{\partial N_{Ay}}{\partial y} + \frac{\partial N_{Az}}{\partial z} = R_A \tag{C.4}$$

Cylindrical:

$$\frac{\partial c_A}{\partial t} + \frac{1}{r}\frac{\partial (rN_{Ar})}{\partial r} + \frac{1}{r}\frac{\partial N_{A\theta}}{\partial \theta} + \frac{\partial N_{Az}}{\partial z} = R_A \tag{C.5}$$

Spherical:

$$\frac{\partial c_A}{\partial t} + \frac{1}{r^2}\frac{\partial (r^2N_{Ar})}{\partial r} + \frac{1}{r\sin\theta}\frac{\partial (N_{A\theta}\sin\theta)}{\partial \theta} + \frac{1}{r\sin\theta}\frac{\partial N_{A\phi}}{\partial \phi} = R_A \tag{C.6}$$

C.1.2 Components of the Diffusive Flux Vector in Different Coordinate Systems (Fick's Law)

Rectangular:

$$J^*_{Ax} = -cD_{AB}\frac{\partial x_A}{\partial x}\ J^*_{Ay} = -cD_{AB}\frac{\partial x_A}{\partial y}\ J^*_{Az} = -cD_{AB}\frac{\partial x_A}{\partial z} \tag{C.7}$$

Cylindrical:

$$J^*_{Ar} = -cD_{AB}\frac{\partial x_A}{\partial r}\ J^*_{A\theta} = -cD_{AB}\frac{1}{r}\frac{\partial x_A}{\partial \theta}\ J^*_{Az} = -cD_{AB}\frac{\partial x_A}{\partial z} \tag{C.8}$$

Spherical:

$$J^*_{Ar} = -cD_{AB}\frac{\partial x_A}{\partial r}\ J^*_{A\theta} = -cD_{AB}\frac{1}{r}\frac{\partial x_A}{\partial \theta}\ J^*_{Az} = -cD_{AB}\frac{1}{r\sin\theta}\frac{\partial x_A}{\partial z} \tag{C.9}$$

C.1.3 Species A Continuity Equation for Constant C and D_A in a Dilute System (Convective–Diffusion Equation)

$$\frac{\partial c_A}{\partial t} + \mathbf{v}\cdot\nabla c_A = D_A\nabla^2 c_A + R_A \tag{C.10}$$

Rectangular:

$$\frac{\partial c_A}{\partial t} + v_x\frac{\partial c_A}{\partial x} + v_y\frac{\partial c_A}{\partial y} + v_z\frac{\partial c_A}{\partial z} = D_A\left(\frac{\partial^2 c_A}{\partial x^2} + \frac{\partial^2 c_A}{\partial y^2} + \frac{\partial^2 c_A}{\partial z^2}\right) + R_A \tag{C.11}$$

Cylindrical:

$$\frac{\partial c_A}{\partial t} + v_r\frac{\partial c_A}{\partial r} + \frac{v_\theta}{r}\frac{\partial c_A}{\partial \theta} + v_z\frac{\partial c_A}{\partial z} = D_{AB}\left[\frac{1}{r}\frac{\partial}{\partial r}\left(r\frac{\partial c_A}{\partial r}\right) + \frac{1}{r^2}\frac{\partial^2 c_A}{\partial \theta^2} + \frac{\partial^2 c_A}{\partial z^2}\right] + R_A \tag{C.12}$$

Spherical:

$$\frac{\partial c_A}{\partial t} + v_r\frac{\partial c_A}{\partial r} + \frac{v_\theta}{r}\frac{\partial c_A}{\partial \theta} + \frac{v_\varphi}{r\sin\theta}\frac{\partial c_A}{\partial \varphi}$$

$$= D_{AB}\left[\frac{1}{r^2}\frac{\partial}{\partial r}\left(r^2\frac{\partial c_A}{\partial r}\right) + \frac{1}{r^2\sin\theta}\frac{\partial}{\partial \theta}\left(\sin\theta\frac{\partial c_A}{\partial \theta}\right) + \frac{1}{r^2\sin^2\theta}\frac{\partial^2 c_A}{\partial \varphi^2}\right] + R_A \tag{C.13}$$

C.1.4 Components of the Diffusive Flux Vector Constant C and D_A in a Dilute System

Rectangular:

$$N_{Ax} = -D_A \frac{\partial c_A}{\partial x} \quad N_{Ay} = -D_A \frac{\partial c_A}{\partial y} \quad N_{Az} = -D_A \frac{\partial c_A}{\partial z} \tag{C.14}$$

Cylindrical:

$$N_{Ar} = -D_A \frac{\partial c_A}{\partial r} \quad N_{A\theta} = -D_A \frac{1}{r}\frac{\partial c_A}{\partial \theta} \quad N_{Az} = -D_A \frac{\partial c_A}{\partial z} \tag{C.15}$$

Spherical:

$$N_{Ar} = -D_A \frac{\partial c_A}{\partial r} \quad N_{A\theta} = -D_A \frac{1}{r}\frac{\partial c_A}{\partial \theta} \quad N_{A\varphi} = -D_A \frac{1}{r\sin\theta}\frac{\partial c_A}{\partial \varphi} \tag{C.16}$$

Appendix D: Bessel Functions

The ordinary differential equation

$$t^2y''(t) + ty'(t) + (t^2 - \nu^2)y(t) = 0 \tag{D.1}$$

where ν is a known constant. This is called Bessel's equation. Solutions of this equation have been found in terms of power series and their use has become widespread and standard. The general solution of Bessel's equation is

$$y(t) = AJ_\nu(t) + BY_\nu(t) \tag{D.2}$$

where J_ν and Y_ν are the Bessel functions of order ν of the first and second kind, respectively.

Bessel functions have the following properties:

1. They are oscillating functions.
2. At $t = 0$ they satisfy

$$J_0(0) = 1 \tag{D.3}$$

$$J_\nu(0) = 0, \ \nu > 0 \tag{D.4}$$

$$Y_\nu(0) = -\infty \tag{D.5}$$

3. They all tend asymptotically to zero at large values of the argument:

$$\lim_{t\to\infty} J_\nu(t) = 0 \tag{D.6}$$

$$\lim_{t\to\infty} Y_\nu(t) = 0 \tag{D.7}$$

Figures D.1 and D.2 show the behavior of some Bessel functions.

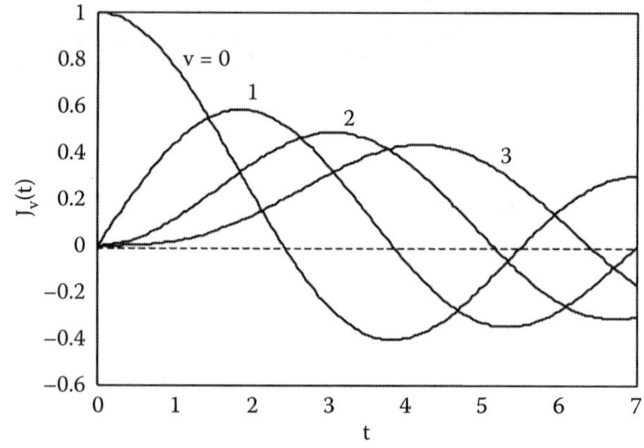

FIGURE D.1 Bessel functions of the first kind and integer order.

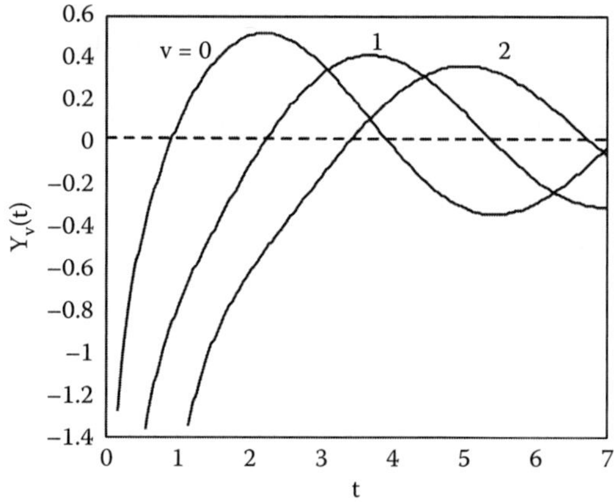

FIGURE D.2 Bessel functions of the second kind and integer order.

Index

A

B

C